Elements

HYDROGEN
AND THE NOBLE GASES

H

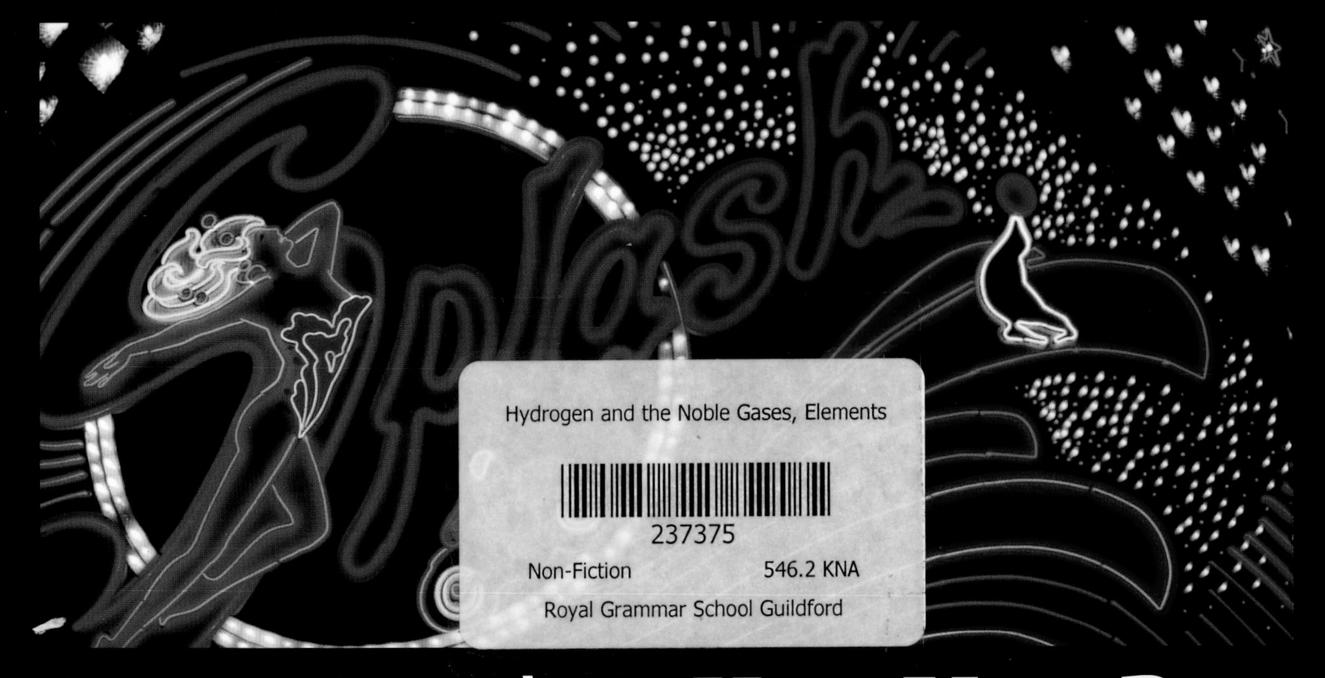

He Ne Ar Kr Xe Rn

Atlantic Europe Publishing

How to use this book

This book has been carefully developed to help you understand the chemistry of the elements. In it you will find a systematic and comprehensive coverage of the basic qualities of each element. Each two-page entry contains information at various levels of technical content and language, along with definitions of useful technical terms, as shown in the thumbnail diagram to the right. There is a comprehensive glossary of technical terms at the back of the book, along with an extensive index, key facts, an explanation of the Periodic Table, and a description of how to interpret chemical equations.

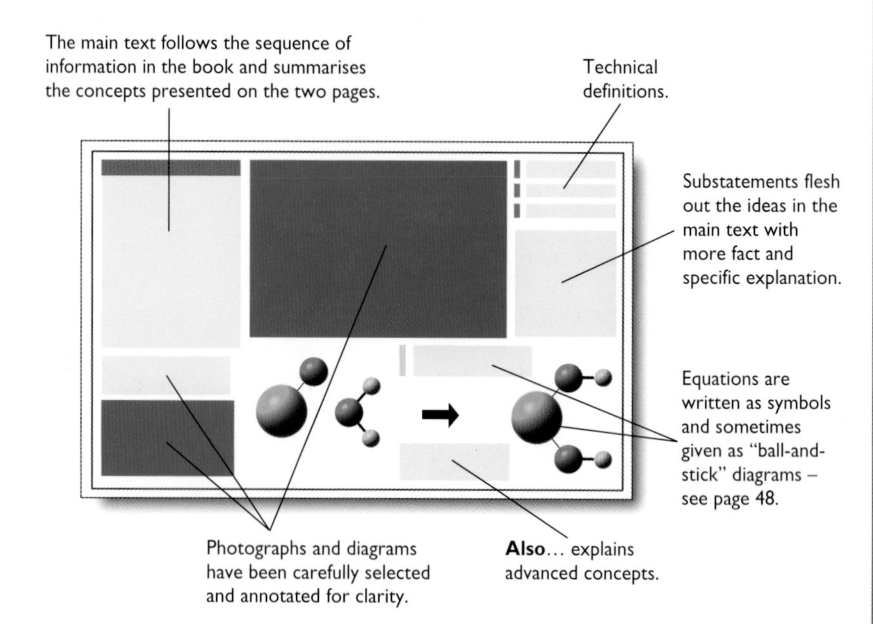

An Atlantic Europe Publishing Book

Author
Brian Knapp, BSc, PhD
Project consultant
Keith B. Walshaw, MA, BSc, DPhil (Head of Chemistry, Leighton Park School)
Industrial consultant
Jack Brettle, BSc, PhD (Chief Research Scientist, Pilkington plc)
Art Director
Duncan McCrae, BSc
Editor
Elizabeth Walker, BA
Special photography
Ian Gledhill
Illustrations
David Woodroffe and David Hardy
Electronic page make-up
Julie James Graphic Design
Designed and produced by
EARTHSCAPE EDITIONS
Print consultants
Landmark Production Consultants Ltd
Reproduced by
Leo Reprographics
Printed and bound by
Paramount Printing Company Ltd

First published in 1996 by
Atlantic Europe Publishing Company Limited, Greys Court Farm, Greys Court, Henley-on-Thames, Oxon, RG9 4PG, UK.

Suggested cataloguing location
Knapp, Brian
Hydrogen and the noble gases
ISBN 1 869860 79 9
– *Elements* series
540

Acknowledgements
The publishers would like to thank the following for their kind help and advice: *ICI (UK)*.

Picture credits
All photographs are from the **Earthscape Editions** photolibrary except the following:
(c=centre t=top b=bottom l=left r-right)
courtesy of **ICI** 14bl, 16; **Mary Evans Picture Library** 38br and **ZEFA** 41, 42br.

Front cover: Collecting hydrogen over water.
Title page: Neon produces a reddish orange colour when excited electrically and so is used in some lighting.

This product is manufactured from sustainable managed forests. For every tree cut down at least one more is planted.

The demonstrations described or illustrated in this book are not for replication. The Publisher cannot accept any responsibility for any accidents or injuries that may result from conducting the experiments described or illustrated in this book.

Contents

Introduction

An element is a substance that cannot be broken down into a simpler substance by any known means. Each of the 92 naturally occurring elements is therefore one of the fundamental materials from which everything in the Universe is made. This book is about hydrogen and the noble gases.

Hydrogen

Hydrogen (chemical symbol H), named after the Greek word for "water-forming", is the most abundant element in the Universe. In fact more than nine-tenths of all atoms are hydrogen atoms. It is the stuff of stars, of cold "empty" space and of the Earth. Hydrogen is found everywhere and in most compounds. It is the most universal element.

Hydrogen is also the simplest element in existence. Its atom contains just one proton in its core, and only one electron is associated with it. As a result, hydrogen is a tiny atom. But its simplicity explains why it is so universal: it is the building block from which other elements are made.

Although it is so universal, hydrogen is rarely found as a gas on Earth. It makes up only three-quarters of 1% of the mass of the planet. This is because hydrogen molecules (combinations of two hydrogen atoms, described by the symbol H_2) weigh so little that they can escape from the Earth's atmosphere. The only free hydrogen that survives is in pockets deep underground, part of the decay process that also forms oil and natural gas.

Hydrogen is a gas that easily catches fire and can be explosive. Yet it is made in huge quantities each year. About half of it is used to make ammonia (a compound of nitrogen and hydrogen), which is the basis for many fertilisers. About one-third of the hydrogen produced is used in refining metals. Another main use of hydrogen is to make a liquid called methanol, one of the starting materials in making artificial fibres.

The noble gases

Helium (symbol He) is another element with a simple structure. It is the second most abundant element in the Universe, and yet its presence was not suspected until relatively recently. It is the only element ever to have been identified in space before it was found on Earth. Scientists noticed a mysterious element they could not identify in the light shining from stars. Eventually this was identified as helium.

Helium, like hydrogen, is lighter than air. But it is an inert gas, which means it rarely reacts and cannot burn. Being inert, it is not found in any compounds on Earth, which is why it took so long to identify. But being an inert gas also means it can be used in places where hydrogen would be dangerous. Thus, helium, rather than hydrogen, fills the floating balloons seen at funfairs.

Helium is one of a group of inert, or noble, gases, whose physical and chemical properties are closely related. The other noble gases, which together make about 1% of the Earth's atmosphere, are neon, argon, krypton, xenon and radon. All other elements react to form stable substances. The noble gases are unreactive because they each have just enough electrons to be perfectly stable on their own, without needing to react with other elements.

▶ The immense heat of the Sun is produced by burning hydrogen.

Hydrogen in the stars

The Sun is made mainly of hydrogen. Burning hydrogen is the source of sunlight and all other starlight.

The Sun's immense gravitational force pulls its hydrogen atoms together, and by a process called fusion, helium is formed and energy released. Thus, hydrogen is used as the fuel in the Sun.

Fusion is the most powerful reaction in the Universe. People have copied it to make thermonuclear (hydrogen) bombs. The conversion of hydrogen makes the Sun intensely hot and releases vast amounts of energy to space as radiation.

A small fraction of this radiation reaches Earth as sunlight, the visible rays of the spectrum, but more radiation reaches the Earth as ultraviolet light, infra-red light and a variety of other wavelengths of radiation that we cannot actually see. In this way nuclear fusion reactions in the Sun make possible life on Earth.

Fusion: combining hydrogen atoms

Nuclear fusion is the forcing together, or fusing, of two atoms to produce an even heavier atom. When fusion occurs there is an enormous release of energy.

Fusion has proved to be one of the most difficult processes for scientists to achieve artificially, and yet it is one of the most common features of the Universe. The reason fusion is so difficult is that enormously high temperatures and great pressures are needed. In the stars both of these conditions are naturally present. So getting fusion going on Earth is nothing short of making stars.

Pollution-free radiation

Fusion does not create any radioactive pollution. It is not a chain reaction like fission and so cannot run out of control and cause a massive explosion. Thus the present generation of nuclear reactors is just a temporary step on the way to the goal of clean, safe nuclear energy. This is why countries need to pursue nuclear technology, so that they do not fall behind and can really benefit from the development of fusion power when it comes.

Phenomenal energy from hydrogen

Deep in the interior of any star the temperature is high enough (10,000,000°C) and the density great enough (30 g/cu cm, ten times the density of rocks on the Earth's surface) for a nuclear fusion reaction to occur.

Stars have a central core composed of helium (made from hydrogen), surrounded by a shell where hydrogen is converted to new helium, and an outer shell composed mostly of hydrogen (which will eventually be used as fuel). This outer shell determines the life of the star. When all of its fuel is used up, the star will die.

fuel: a concentrated form of chemical energy. The main sources of fuels (called fossil fuels because they were formed by geological processes) are coal, crude oil and natural gas. Products include methane, propane and gasoline. The fuel for stars and space vehicles is hydrogen.

fusion: combining atoms to form a heavier atom.

radiation: the exchange of energy with the surroundings through the transmission of waves or particles of energy. Radiation is a form of energy transfer that can happen through space; no intervening medium is required (as would be the case for conduction and convection).

thermonuclear reactions: reactions that occur within atoms due to fusion, releasing an immensely concentrated amount of energy.

An endless supply of cheap energy

Scientists know that if they can use fusion to make energy they can use hydrogen from ordinary water (in a form called heavy water). This supply will last for billions of years.

The main problem is that, because a nuclear reactor can only be a tiny faction of the size of a star, the size (and gravity) must be compensated for by creating even higher temperatures than exist in stars. This is what scientists are working on today and why it may take to the middle of next century to achieve it.

Also...

In the inner region of a star four hydrogen atoms are converted into one helium atom. During this process about one-third of the mass is lost and converted into energy.

Einstein's famous equation $E = mc^2$, where E is energy, m is mass and c is the velocity of light, shows that when mass is lost it is converted into phenomenal amounts of energy. A nuclear reaction inside a star delivers energy from hydrogen 60 million times more effectively than chemical reactions such as the burning of hydrogen in oxygen. The Sun converts 600 million tonnes of hydrogen each second, creating 400 million tonnes of helium. Part of this energy occurs as visible light rays, accounting for the brightly glowing ball that we see as the Sun.

How hydrogen bonds

Hydrogen is found in many compounds. The tiny hydrogen atoms bond to atoms of nitrogen, oxygen and fluorine in a very special way, like a special kind of glue. This is called hydrogen bonding.

Hydrogen bonding determines the way water molecules lock together to make ice: in this solid, hydrogen bonds hold the molecules apart in an open structure, like a honeycomb.

In the case of ice, the framework is rigid and so strong that ice a few centimetres thick can hold the weight of a person. However, hydrogen bonds also allow cellulose, the tissue of all plant life, to be a rigid yet *flexible* framework.

Hydrogen bonds are at work in our bodies as well. They make the springy but tough sinews of our tendons. In a similar way, hydrogen bonds provide the vital zipper that connects the two strands of DNA molecules, the blueprint of life (see page 9). The weak hydrogen bonds can easily split apart, allowing the strands to duplicate, then re-form.

▲ Ice has an open structure that makes it less dense than liquid water.

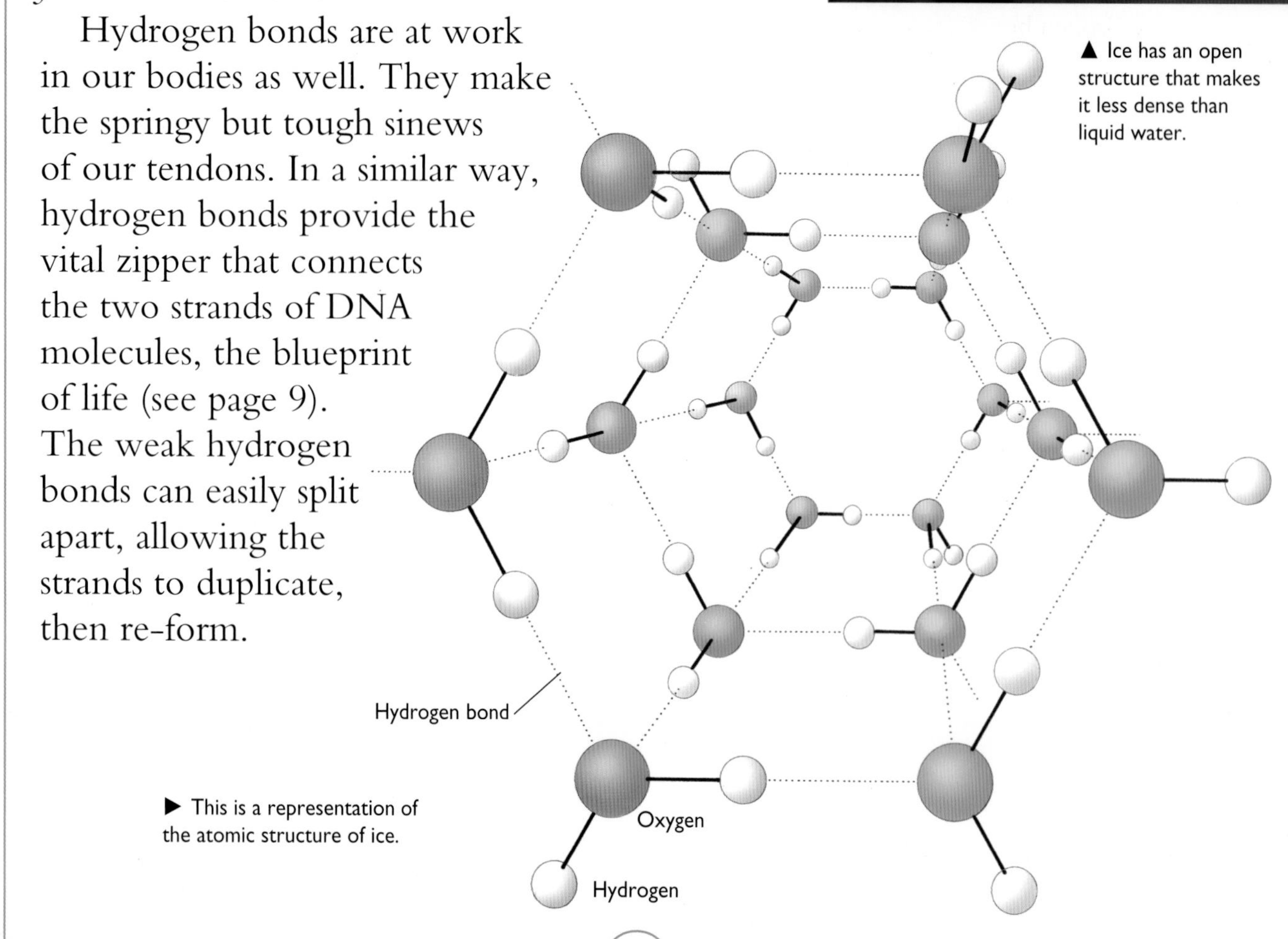

▶ This is a representation of the atomic structure of ice.

Hydrogen bonds and the genetic code

Life depends on two fundamental blueprint acids, deoxyribonucleic acid (DNA) and ribonucleic acid (RNA). These acids carry the genetic code from one generation to another.

Each DNA molecule is made of a number of short segments or units that are connected together forming two long, twisting coils known as a double helix.

The coils have to be held together firmly and yet be able to split up easily and re-form as part of the process of reproduction. The key to this is that the two coils are held together by hydrogen bonds.

During reproduction, the two coils have to unzip and each has to copy itself. The unzipping is easy because hydrogen bonds are relatively weak. Each separated strand then attracts new molecules, in a very specific order, which bond to the coils by hydrogen bonds until a new pair of coils has been produced.

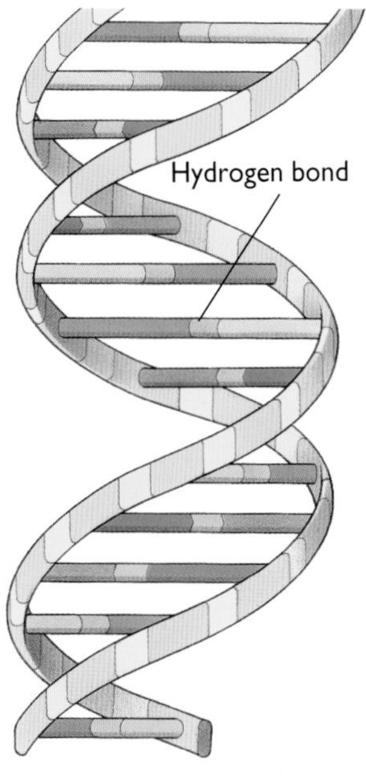

▲ The double helix structure of DNA.

bond: chemical bonding is either a transfer or sharing of electrons by two or more atoms. There are a number of types of chemical bond, some very strong (such as covalent bonds), others weak (such as hydrogen bonds). Chemical bonds form because the linked molecule is more stable than the unlinked atoms from which it formed. For example, the hydrogen molecule (H_2) is more stable than single atoms of hydrogen, which is why hydrogen gas is always found as molecules of two hydrogen atoms.

hydrogen bond: a type of attractive force that holds one molecule to another. It is one of the weaker forms of intermolecular attractive force.

Also...

Hydrogen bonds occur because of the way that hydrogen is attracted to some other atoms, such as oxygen. The oxygen pulls at the hydrogen in such a way that the side of the hydrogen atom facing away from the oxygen is left with a slight positive charge. This allows the hydrogen to attract other molecules that are negatively charged and to form a link, or bond. In this sense, hydrogen bonding is rather like the way a balloon sticks to a sweater after it has been rubbed vigorously. It can be pulled off and stuck back at will, a very flexible, if not very strong, electrical attraction.

Hydrogen bonds are very weak compared to other bonds that hold atoms together as molecules. They are nowhere as strong as the (ionic) bonds in salt or the bonds in the metal iron or a mineral like silica. But these other very strong types of bond seldom allow materials to be flexible, or to break up and re-form easily.

Hydrogen bonding also has a special effect on H_2O – water. Water should actually be a gas since its molecules are relatively simple. (The more complex the molecule, the more attracted one molecule is to another and the less likely it is to be a gas.) But hydrogen bonding between the molecules keeps water as a liquid, unless a lot of heat energy is applied.

◀ This spider's web is held together by hydrogen bonding. The silk in the web is made of proteins stuck together by hydrogen bonds so that the silk has flexibility yet has as much strength as a strand of steel the same size.

Hydrogen in water

Water is the most common chemical compound on the surface of the Earth, covering over two-thirds of the Earth's surface. People are also about two-thirds water.

Water is a compound of hydrogen and oxygen. Although we are so familiar with the way water behaves that we take it as normal, water, in fact, behaves differently from most other compounds. This is due to the strong attraction that each water molecule has for others. Water shrinks as it is cooled (down to 4°C) and then expands as it is cooled further still (to 0°C). It is also the only common substance that swells as it freezes.

All these unusual properties are due to the special way that the hydrogen atoms link water molecules together, the property called hydrogen bonding (see pages 8 and 9).

The atoms that make up water molecules are bound together very much more firmly than are the molecules to each other. Enormous amounts of energy are needed to break them apart. Heating, for example, only changes the physical state of water, converting solid ice to liquid water and then to steam. Only electrical energy will *break* bonds within the molecules.

The process of decomposing water (breaking it apart) is called electrolysis. The demonstration on these pages uses electrolysis to collect hydrogen and oxygen gases separately. The equipment is called Hoffman's voltameter.

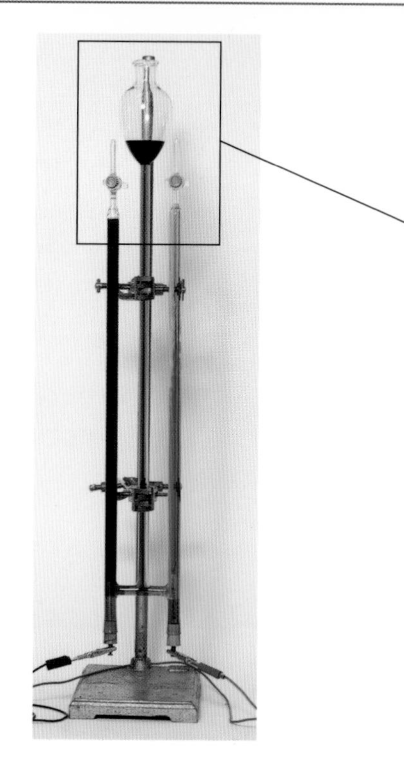

▲ This is Hoffman's voltameter. It consists of three connected glass tubes with electrodes in the bottom of the outer two. The electrodes are connected to a source of direct current (dc) electricity.

Also...

Water (H_2O) is one of the most common substances to contain hydrogen, yet it has some unusual properties. Only a few water molecules (about one in 500 million) break up into ions at room temperature. Because it has so few ions, *pure* (sometimes called deionised) water is a poor conductor of electricity (but see the discussion of how electrolysis works on page 11).

Hydrogen ions can exist only in the presence of water, which explains why an acid (see page 18) behaves as an acid only when in water and why it does not react as solid crystals. Thus baking powder (which contains tartaric acid) only produces bubbles of hydrogen when mixed with water.

bond: chemical bonding is either a transfer or sharing of electrons by two or more atoms. There are a number of types of chemical bond, some very strong (such as covalent bonds), others weak (such as hydrogen bonds). Chemical bonds form because the linked molecule is more stable than the unlinked atoms from which it formed. For example, the hydrogen molecule (H_2) is more stable than single atoms of hydrogen, which is why hydrogen gas is always found as molecules of two hydrogen atoms.

covalent bond: the most common form of strong chemical bonding, which occurs when two atoms *share* electrons.

dissociate: to break apart. In the case of acids it means to break up forming hydrogen ions. This is an example of ionisation. Strong acids dissociate completely. Weak acids are not completely ionised and a solution of a weak acid has a relatively low concentration of hydrogen ions.

ion: an atom, or group of atoms, that has gained or lost one or more electrons and so developed an electrical charge.

◀ This picture shows a detail of the Hoffman's voltameter, some time after the power has been applied. The indicator measures the amount of hydrogen ions in the water. The hydrogen ions have been converted to hydrogen molecules and form a gas in one tube (above the blue-stained water). The oxygen forms in the tube where the indicator has turned the water red.

Two volumes of hydrogen gas collect for every one volume of oxygen gas, as expected from the formula for water: H_2O. The relative volume is also shown by the equation at the foot of the page.

How electrolysis works

Pure water is a poor conductor of electricity. But water with impurities in it does conduct. Dilute solutions of acids, for example, are excellent conductors of electricity. Dilute sulphuric acid is used in vehicle batteries.

If two electrodes are dipped in a dilute acid solution, the electrical energy will break apart the hydrogen particles (ions) from the hydroxyl ions (pairs of hydrogen and oxygen particles). Because opposite charges attract, positively charged hydrogen ions drift through the dilute acid to the negative electrode. There they form hydrogen gas. Negatively charged hydroxyl ions drift to the positive electrode where oxygen gas forms.

The production of hydrogen gas from water comes at a price. Large amounts of electricity have to be used. However, the energy is not wasted, but simply stored in the hydrogen gas as chemical energy. This energy makes hydrogen gas very inflammable and even explosive if mixed with air or oxygen.

EQUATION: Dissociation of water

Water ➪ hydrogen + oxygen

$2H_2O(l) \Rightarrow 2H_2(g) + O_2(g)$

Preparing hydrogen gas

Hydrogen gas can be produced in the laboratory by reacting a dilute acid with a metal. The demonstration below uses dilute hydrochloric acid and zinc. Dilute copper sulphate has been added to the hydrochloric acid to speed up the reaction; this is what makes the acid green. The copper does not take part in the reaction; it is a catalyst.

What the apparatus does

Hydrogen is produced by the reaction in the flask. The gas is then led through tubing into a water bath containing a gas jar supported on a beehive shelf. This is the standard way of collecting gases that are not soluble in water. As the gas bubbles through it, the water is displaced but the gas is contained.

At the end of the experiment a cover slip can be placed on the open end of the gas jar while it is still under water, thus sealing the gas in the jar.

catalyst: a substance that speeds up a chemical reaction but itself remains unaltered at the end of the reaction.

EQUATION: Laboratory production of hydrogen

Hydrochloric acid + zinc ➪ zinc chloride + hydrogen

$2HCl(aq)$ + $Zn(s)$ ➪ $ZnCl_2(aq)$ + $H_2(g)$

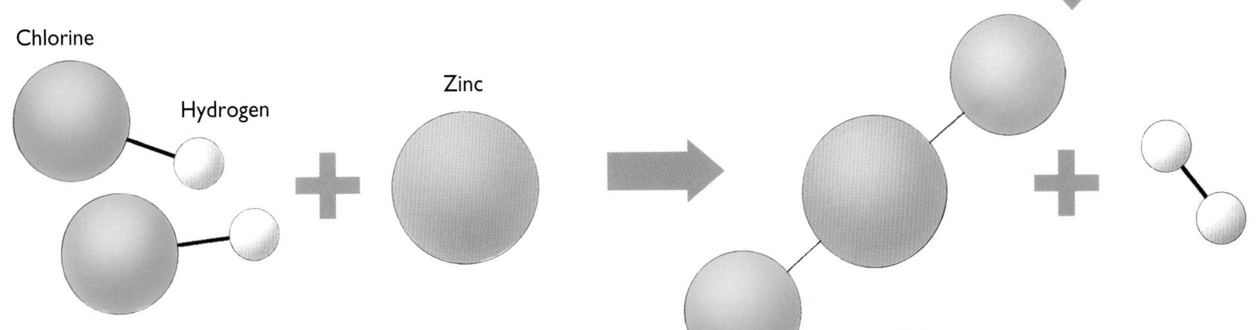

Also...

Hydrogen gas burns with a colourless flame and at a high temperature, and tends to melt the glass around the end of the tube as it burns. The orange-red colour of the flame you see is due to the presence of sodium in the glass; the sodium colours the flame.

Inflammable hydrogen

If hydrogen is not collected in the gas jar, it will seep out of the open tubing. In this picture the gas has been set alight, showing that hydrogen is inflammable.

The small size of the flame shows that hydrogen is not produced at a particularly fast rate, but the flame is steady, showing how the copper sulphate catalyst plays a useful part in making the gas production constant. (Hydrogen actually burns with a colourless flame; the yellow colour is from some of the sodium in the glass, released as the flame heats the glassware.)

Manufacturing hydrogen

There is a great need for hydrogen gas in industry. It can be made on a large scale in a number of ways. Essentially, the raw materials are water and hydrocarbons (mixtures of compounds of hydrogen and carbon, such as crude oil). Hydrogen gas can be produced by reacting water (as steam) on white-hot coke. Steam on burning crude oil can also be used, with the aid of a nickel catalyst.

Hydrogen is also produced as a byproduct during the cracking of crude oil or the processing of brine in a diaphragm cell.

Also...

New developments in hydrogen production have occurred, allowing hydrogen to be produced on quite a different scale to that in the past. A small chamber is used as an electrolytic cell. High frequency, high voltage electricity is applied across the cell. This causes the hydrogen to break away from the oxygen with almost no heat given out.

The preparation of substantial amounts of hydrogen on this small scale provides the possibility of making a portable hydrogen producer that could be used to provide the fuel for cars. When this process is perfected, filling up the tank of a vehicle with fuel could be as easy as taking a hose from the nearest water tap!

Hydrogen from hydrocarbons

In a cracking plant the heavier products of an oil refinery are split up into smaller ones. A variety of methods are used, including heat, pressure and a vacuum.

The most important petrochemical production process is "steam cracking". This process takes place at temperatures of about 800°C.

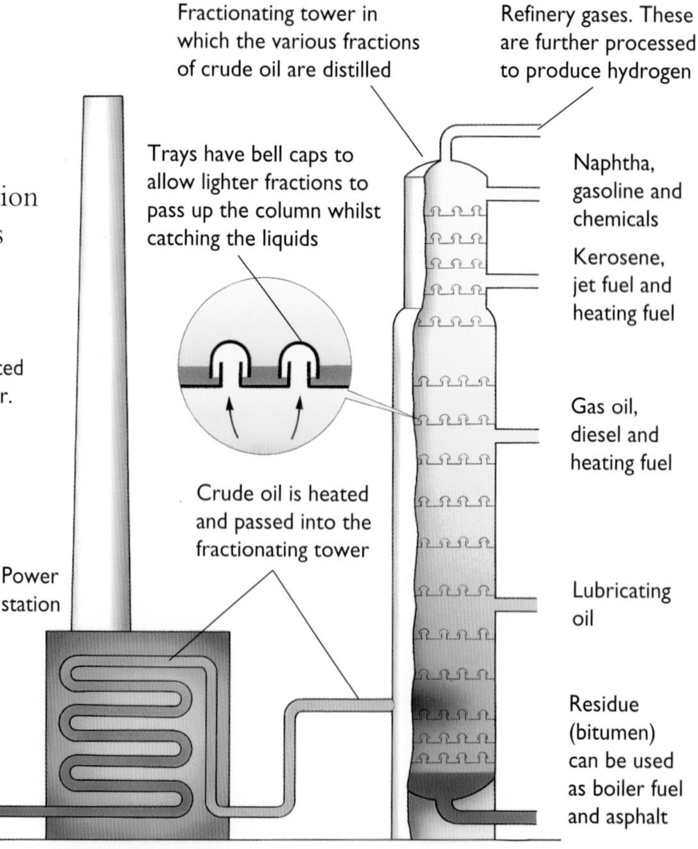

▶ This diagram shows the nature of a fractionating tower. Hydrogen is collected with other gases at the top of the tower.

▲ This is part of a petrochemical plant. The slim towers are a cracking unit in which petrochemical fractions are separated and hydrogen given off as a byproduct.

Production of hydrogen during cracking of petroleum fractions

Crude oil ⇨ octane + ethene + hydrogen

$C_{12}H_{26}(aq)$ ⇨ $C_8H_{18}(g)$ + $2C_2H_4(g)$ + $2H_2(g)$

Manufacture of hydrogen from brine

This method uses an electrical method for collecting hydrogen as part of a complicated reaction that yields a number of other important chemicals and uses cheap raw materials.

The reactions take place in an electrolytic cell, in which the two halves of the cell are separated by a semipermeable membrane or diaphragm. The pores of the semipermeable membrane are designed to allow only sodium ions to pass through.

Brine, containing sodium and chloride ions, is pumped into one half of the cell. An electrical current causes the sodium ions to be attracted through the semipermeable membrane. The used-up brine is pumped out of the cell and more concentrated brine is pumped in.

A supply of sodium hydroxide is pumped through the left hand side of the cell. The sodium ions form more sodium hydroxide, thus increasing the concentration of the sodium hydroxide. At the same time, surplus hydrogen ions in the water are converted to hydrogen gas, which bubbles out of the solution and is collected.

The diaphragm also prevents the hydrogen and chlorine gases from mixing and allows them to be collected separately.

cracking: breaking down complex molecules into simpler components. It is a term particularly used in oil refining.

crude oil: a chemical mixture of petroleum liquids. Crude oil forms the raw material for an oil refinery.

electrolysis: an electrical–chemical process that uses an electric current to cause the break up of a compound and the movement of metal ions in a solution.

semipermeable membrane: a thin membrane of material that acts as a fine sieve, allowing small molecules to pass, but holding large molecules back.

▼ A diagrammatic representation of the electrolysis process for manufacturing sodium hydroxide.

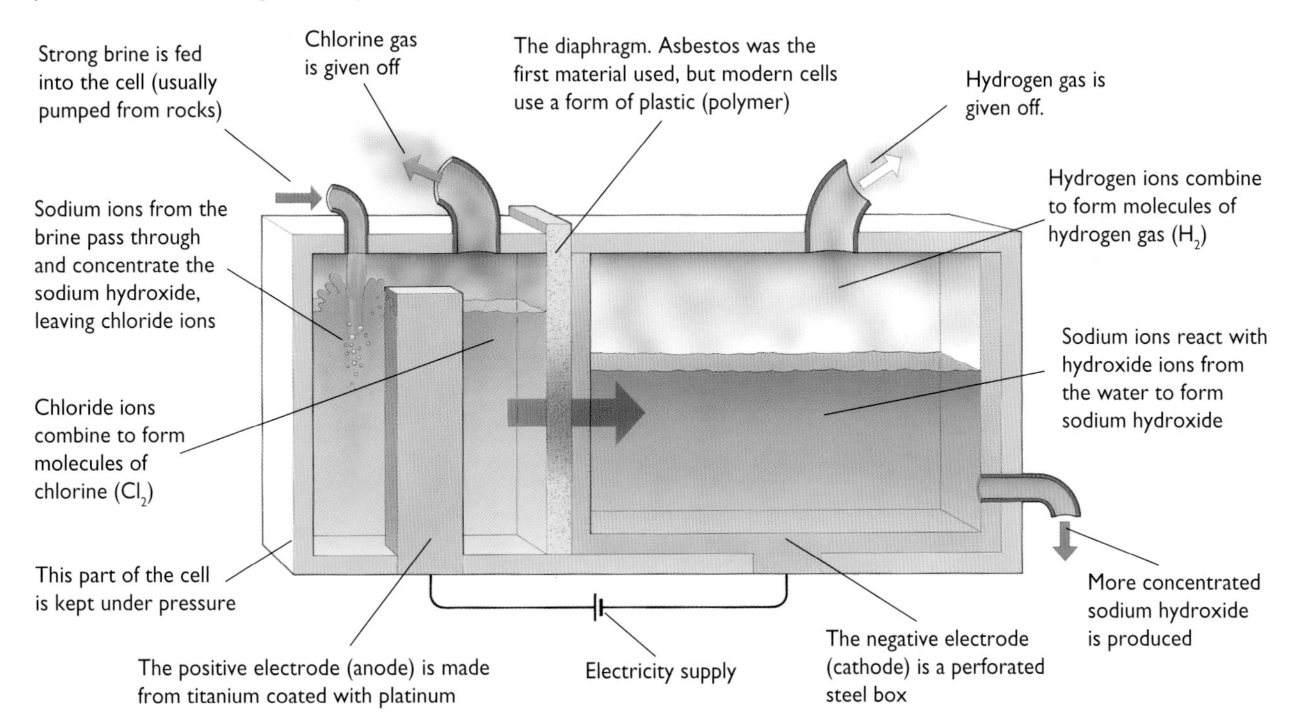

EQUATION: Electrolysis of a salt solution

Sodium chloride + water ⇨ sodium hydroxide + chlorine gas + hydrogen gas

$$2NaCl(aq) + 2H_2O(l) \xrightarrow[\text{electrical energy}]{} 2NaOH(aq) + Cl_2(g) + H_2(g)$$

Hydrogen for ammonia

About a half of all the hydrogen produced in industry is used for making ammonia. Ammonia is one of the foundation materials for the manufacture of artificial fertilisers. These fertilisers, in turn, are vital for boosting the world's food supply.

Ammonia

Ammonia is a compound of nitrogen and hydrogen. It is easy to detect because it has a strong, irritating smell. It is a colourless gas, not actually poisonous, but breathing in its fumes can cause people to stop breathing, so it can have very serious effects.

Ammonia is extremely soluble in water, where it produces an alkaline liquid called ammonium hydroxide. This dilute solution is commonly sold for use as a household disinfectant and cleaning agent under the title "household ammonia".

Making ammonia

The method of making ammonia is called the Haber–Bosch process after its inventors. In the process, hydrogen and nitrogen gases are reacted at high temperatures (up to 600°C) and high pressures (up to 600 atmospheres).

The reaction works best if this is done in the presence of iron, although iron plays no part in the actual reaction (it is a catalyst for this reaction).

The ammonia produced is liquified to separate it from any unreacted gases, and these are then recycled.

◀ This is part of a Haber–Bosch ammonia plant. These pipes are recycling some of the unreacted gases.

Limitations of the Haber–Bosch process

Although the Haber–Bosch process has produced the foundation material for important fertilisers and has thus greatly contributed to the increase in the world's food supply, it consumes large amounts of energy and the equipment, operating at high temperatures and pressures, is expensive to manufacture.

The source of hydrogen for this reaction is mainly petroleum, a nonrenewable fossil fuel. For these reasons, fertilisers are expensive and scientists are looking for a natural way to fix nitrogen for use as a fertiliser, especially for use in the developing world.

alkaline: the opposite of acidic. Alkalis are bases that dissolve, and alkaline materials are called basic materials. Solutions of alkalis have a pH greater than 7.0 because they contain relatively few hydrogen ions.

catalyst: a substance that speeds up a chemical reaction but itself remains unaltered at the end of the reaction.

fixation of nitrogen: the processes that natural organisms, such as bacteria, use to turn the nitrogen of the air into ammonium compounds.

▼ A diagram of the Haber–Bosch process.

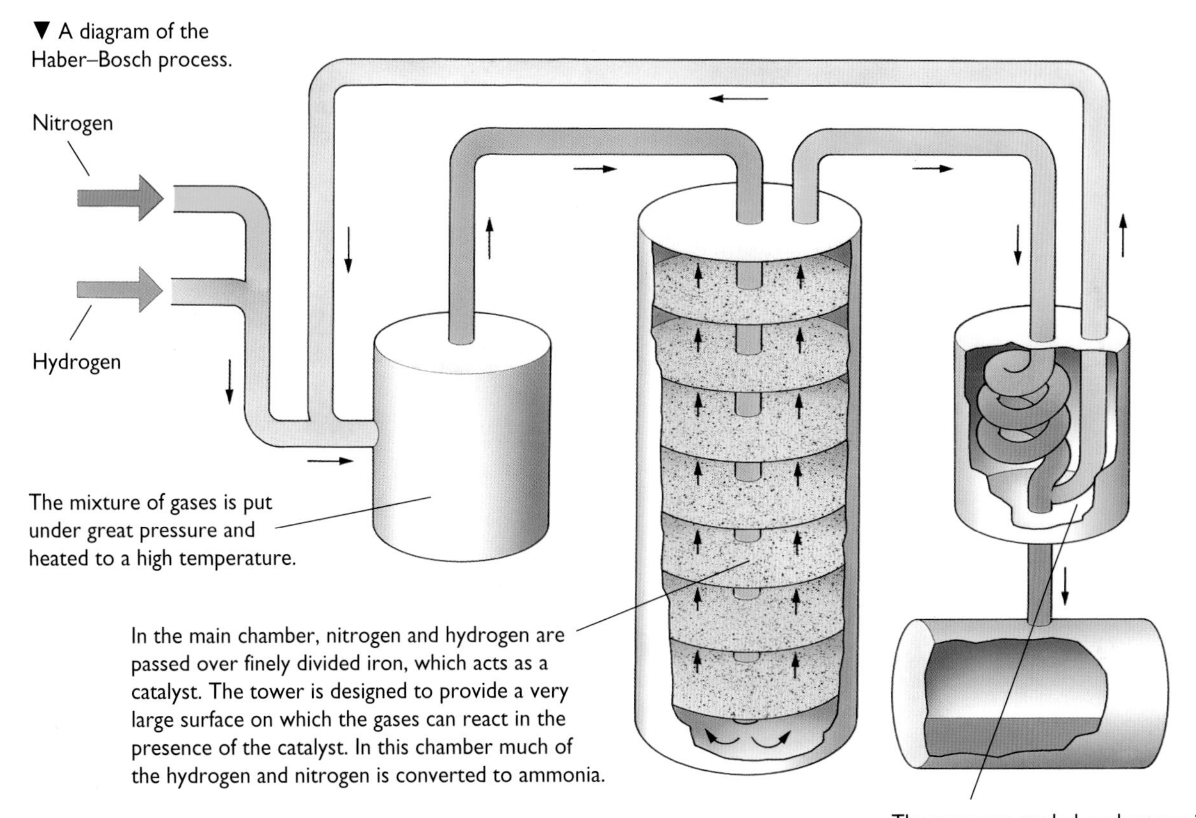

EQUATION: The production of ammonia

Hydrogen + nitrogen ⇨ ammonia

$3H_2(g) + N_2(g) \Rightarrow 2NH_3(g)$

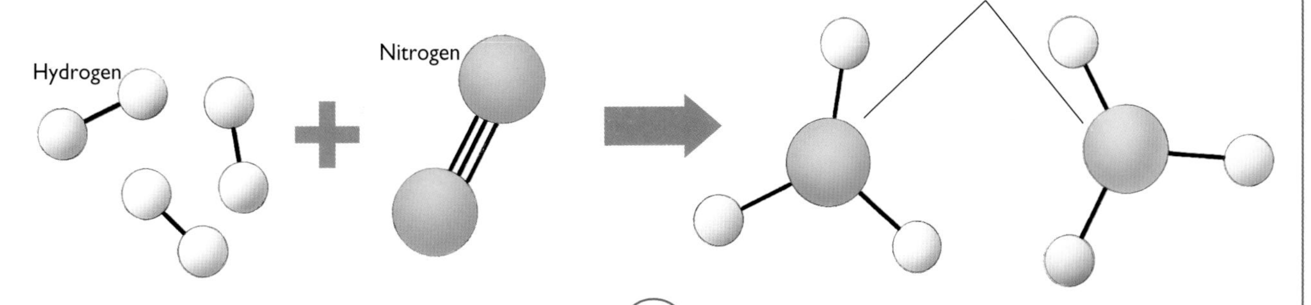

Acids

Acids, whose name comes from *acetum*, the Latin for vinegar, are a large and important group of chemicals. They are usually corrosive and have a sour taste. Acids are grouped into mineral acids, such as sulphuric, nitric and hydrochloric acid, and organic acids (containing carbon), such as tartaric and citric acid.

Mineral acids are produced by many natural processes. For example, acid forms when lightning strikes, as the nitrogen and oxygen in the air are converted into oxides of nitrogen. When they dissolve in the water droplets of a cloud, the result is nitric acid.

Organic acids are produced by living things. One common example is the citric acid produced by citrus fruits, which gives them their sharp taste.

Both mineral and organic acids dissociate to some degree when they are added to water. Acids can also react with some metals to give off hydrogen gas.

Mineral acids

Mineral acids are acids that do not contain any carbon. The most common mineral acids are hydrochloric, sulphuric and nitric acid.

The mineral acids are called strong acids. They all have a hydrogen atom attached to a nonmetal atom.

►▼ When magnesium ribbon is dropped into dilute hydrochloric acid, a reaction takes place in which hydrogen gas and heat are given off, creating the steam shown in the picture. The metal is being corroded by the acid, as shown by the equation below.

EQUATION: The reaction of magnesium with an acid

Magnesium + dilute hydrochloric acid ⇨ magnesium chloride + hydrogen gas

$$Mg(s) + 2HCl(aq) \Rightarrow MgCl_2(aq) + H_2(g)$$

◆

Chlorine

Hydrogen

Magnesium

Strong and weak acids

Some acids dissociate almost completely when they dissolve in water. These are called strong acids, because the solution will contain a large number of hydrogen ions. Sulphuric acid is an example of a strong acid.

Other acids do not dissociate in water readily, resulting in few hydrogen ions in the solution. These are called weak acids, for example, carbonic acid.

▶ Strong acids dissociate in water, making a good electrolyte. This is sulphuric acid being used as an electrolyte in a battery.
The instrument is a hydrometer, a tube with a bulb on the end. The tube contains a small float which is designed so that, when acid is drawn into the tube using the bulb, it will register the changes in density of the battery acid. The changes correspond to the state of charge of the battery. Thus, for example, a fully discharged battery will have a low density, whereas a fully charged battery will have a higher density. The hydrometer is colour coded to make it easy for users to see the state of charge of their battery.

Hydrometer

corrosive: a substance, either an acid or an alkali, that *rapidly* attacks a wide range of other substances.

dissociate: to break apart. In the case of acids it means to break up forming hydrogen ions. This is an example of ionisation. Strong acids dissociate completely. Weak acids are not completely ionised and a solution of a weak acid has a relatively low concentration of hydrogen ions.

mineral acid: an acid that does not contain carbon and that attacks minerals. Hydrochloric, sulphuric and nitric acids are the main mineral acids.

organic acid: an acid containing carbon and hydrogen.

Also...

Do not confuse strong and weak with concentrated and dilute. Strong refers to an acid that will dissociate well in water. Thus both concentrated nitric acid and dilute nitric acid are strong acids.

Dilute and concentrated refer to the proportion of the acid and the solvent. Dilute hydrochloric acid contains far more water than concentrated hydrochloric acid.

Organic acids

All organic acids contain carbon atoms as part of their structure, and most are weak acids. When they are put in water only a small proportion of the acid molecules dissociate at any one time. It is not possible to get a concentrated solution of ions from a weak acid.

▶ Organic acids are weak acids and therefore do not make good electrolytes. This is why the traditional experiment of making a natural battery by putting two electrodes in a lemon (containing citric acid) is quite safe. Very little current will flow.

Testing for acidity (pH)

An important way to measure acidity is to measure the concentration of the hydrogen ions in the solution. The concentration of hydrogen ions is measured by a special scale known as the pH scale. A pH of 0 is strongly acidic, pH 7 is neutral, and pH 14 is strongly alkaline.

The concentration of hydrogen ions can be measured with a pH meter or a chemical indicator. At a specific pH an indicator will change colour. There are many indicators, so that it is possible to find out the pH of a solution by seeing which indicator changes colour.

To make this process as easy as possible, a number of indicator solutions are mixed together to give a Universal Indicator. In the bottle, the Universal Indicator is deep green, but when a few drops are added to a solution of acid or alkali, it changes colour. Deep red indicates a very acidic solution, whereas bright blue indicates a very alkaline solution.

Indicators

Most indicators are complicated organic substances that change colour in different chemical environments, such as acidic or basic solutions. (For example, red cabbage leaves soaked in water are sometimes recommended as indicators in home-based chemistry experiments). The most common acid–base indicator is litmus paper, which is paper treated with a chemical that is red in acidic solutions and blue in basic solutions.

Also... the meaning of pH

The acidity of a solution depends on the relative amounts of hydrogen (H^+) ions and hydroxide (OH^-) ions in the solution. This number varies enormously between very acid and very alkaline, so it is not possible to use a simple (linear) scale to measure the concentration. Instead, the pH scale is a logarithmic scale where each number on the scale represents a ten-fold change in concentration. Thus a solution of pH 5 has ten times the concentration of hydrogen ions as a solution of pH 6. On this scale a pH of 1 is strongly acid, the pH of 14 is strongly alkaline, and a pH of 7 is neutral.

▼ This chart shows the range of pH from a strong acid at pH 1 to a strong alkali at pH 14.

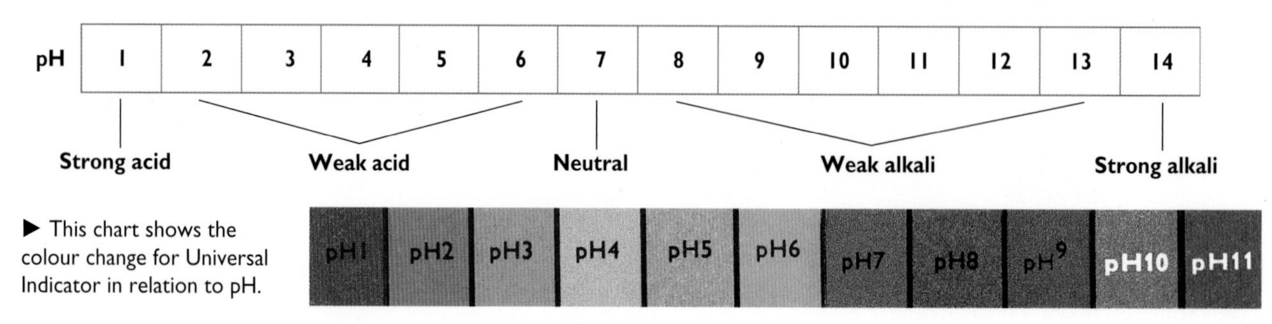

▶ This chart shows the colour change for Universal Indicator in relation to pH.

◀ Litmus is cherry red in an acidic solution and purple-blue in an alkaline solution.

▼ Bromothymol blue is yellow in an acidic solution but blue in an alkaline solution.

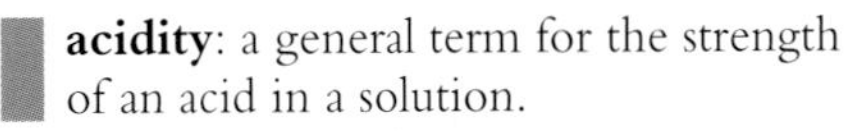

acidity: a general term for the strength of an acid in a solution.

indicator: a substance or mixture of substances that change colour with acidity or alkalinity. Indicators measure the pH of a solution.

ion: an atom, or group of atoms, that has gained or lost one or more electrons and so developed an electrical charge.

pH: a measure of the hydrogen ion concentration in a liquid. Neutral is pH 7.0; numbers greater than this are alkaline, smaller numbers are acidic.

▲ Phenol phthalein is colourless in an acid solution, but turns purple in an alkaline solution.

◀ Methyl orange is red in an acidic solution but yellow in an alkaline solution.

The importance of acid testing

Many chemical processes, the manufacture of food, and successful farming all rely on a careful control of pH.

Environmental scientists measure the pH of falling rainwater. If the pH reading is less than five, then the rain is classed as acid rain. Acid rain is one of the most serious environmental problems of our time.

Hydrochloric acid

Hydrochloric acid is the gas hydrogen chloride (HCl) in water. Hydrochloric acid is a strong, mineral acid, found close to the site of erupting volcanoes. It is also one of the main acids that break down the food in our stomachs (in a concentration of 0.25%).

Most hydrochloric acid is produced as a byproduct when chlorine is reacted with organic compounds as part of the process of making plastics. It is vital for the manufacture of a common plastic, vinyl, or PVC. It is also used in steelworks for cleaning steel, and in extracting metals from their ores.

A mixture of three parts hydrochloric acid and one part nitric acid is called *aqua regia* (royal water) and was traditionally used to dissolve gold (which was known as the "king of metals" by alchemists).

Because hydrochloric acid is readily available as a gas, it can be reacted with a basic gas, such as ammonia, to show that, in the gaseous state, an acid plus a base produces a salt. The demonstration is shown here.

(1)▼ Two colourless gases, one in each cylinder, are brought together. The upper one is ammonia, the lower one hydrogen chloride gas.

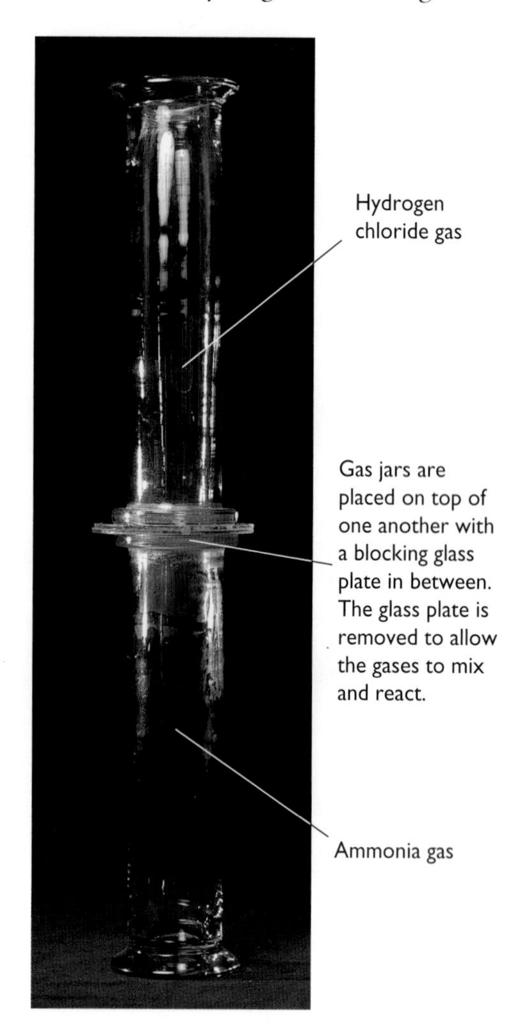

EQUATION: Reaction of hydrochloric acid gas and ammonia

Hydrogen chloride + ammonia ➪ ammonium chloride

$HCl(g)$ + $NH_3(g)$ ➪ $NH_4Cl(s)$

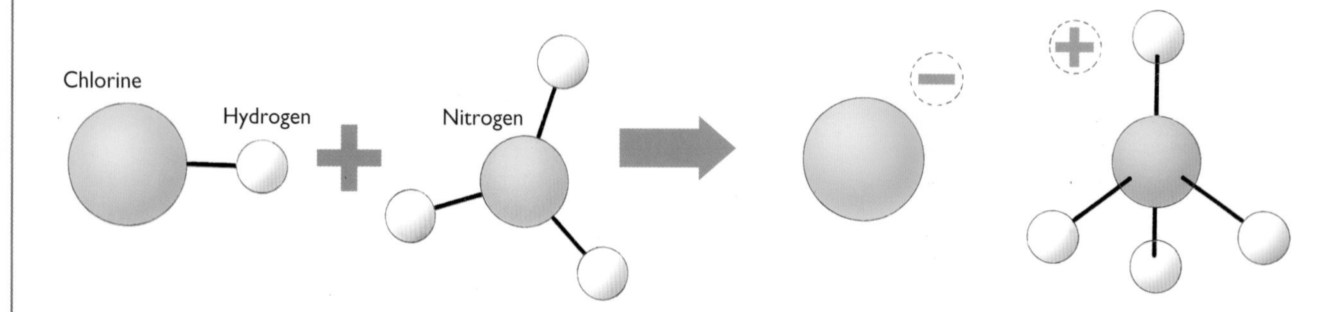

❷▼ Ammonia is a base, so the gases react: an acid plus a base produces a salt.

The reaction produces a "smoke" of white particles of ammonium chloride.

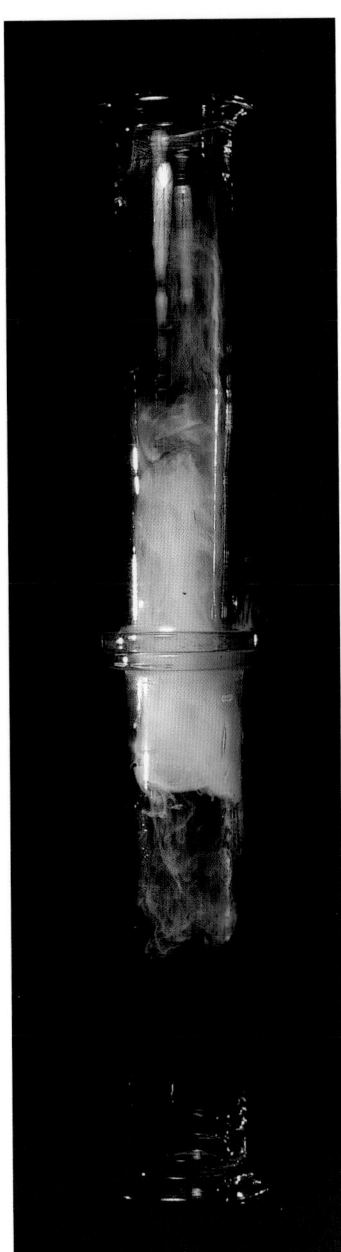

❸▼ Ammonium chloride "smoke" is a suspension of solid particles in a gas, so it begins to settle to the bottom of the gas jar.

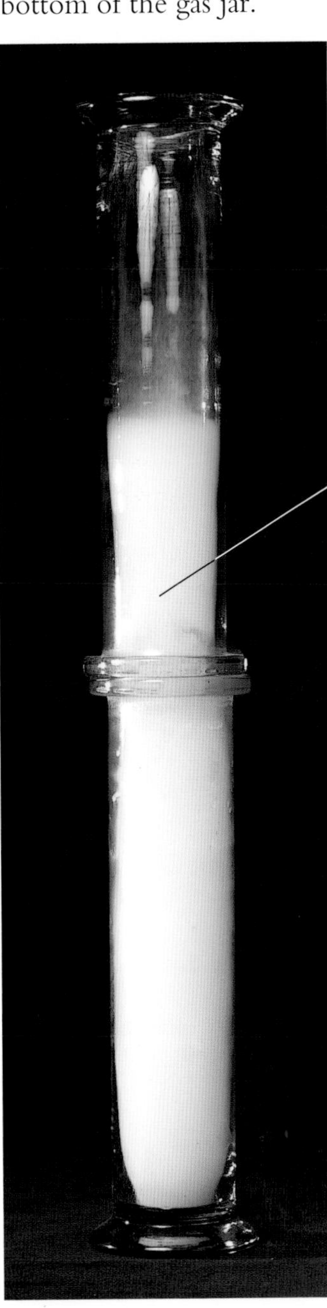

Ammonium chloride "smoke" appears where the hydrogen chloride gas and ammonia gas mix and react.

acid: compounds containing hydrogen which can attack and dissolve many substances. Acids are described as weak or strong, dilute or concentrated, mineral or organic.

alchemy: the traditional "art" of working with chemicals that prevailed through the Middle Ages. One of the main challenges of alchemy was to make gold from lead. Alchemy faded away as scientific chemistry was developed in the 17th century.

base: a compound that may be soapy to the touch and that can react with an acid in water to form a salt and water.

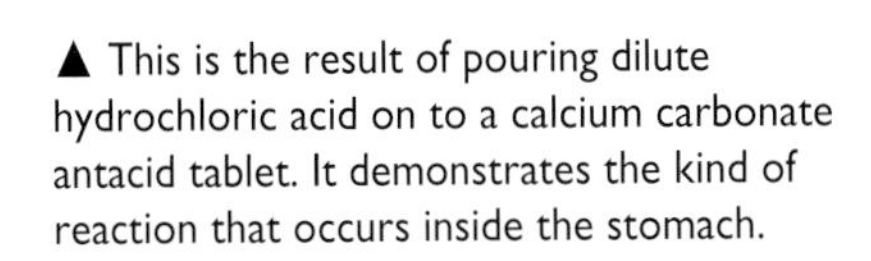

▲ This is the result of pouring dilute hydrochloric acid on to a calcium carbonate antacid tablet. It demonstrates the kind of reaction that occurs inside the stomach.

EQUATION: Reaction of dilute hydrochloric acid and calcium carbonate

Hydrochloric acid + calcium carbonate ⇨ calcium chloride + water + carbon dioxide gas

$2HCl(aq) + CaCO_3(s) \Rightarrow CaCl_2(aq) + H_2O(l) + CO_2(g)$

Sulphuric acid

Sulphuric acid, a colourless, thick, heavy and "oily" liquid is a mineral acid that will react with compounds of most metals. Sulphuric acid is used widely in industry, for example, in the making of fertilisers (superphosphates), paints, synthetic fibres, detergents, dyes, and explosives.

Hot, concentrated sulphuric acid is an oxidising agent. It will react with many materials, even attacking less reactive metals such as copper, as demonstrated opposite.

Here we also use sulphuric acid to demonstrate the ability of acids to remove water from many compounds. For example, if a piece of sugar is placed in sulphuric acid, it becomes a black blob of carbon, all the water having been removed from the sugar molecules.

The effect of removing water is called dehydration. If sulphuric acid touches human skin, the molecules in the skin immediately begin to lose their water. We call this an acid burn. Acid burns are among the most serious accidents that can happen, especially if splashes should get into the eye.

Concentrated sulphuric acid dehydrates sugars

If concentrated sulphuric acid is added to sugar, the sugar dehydrates, going black and losing its water. The reaction produces considerable heat, so water is released as steam. The chemical equation shows that the sulphuric acid remains uncombined; the sugar is simply reduced to carbon (the black material).

As the steam is given off (the reaction is exothermic), bubbles form that cause the carbon to create a substance that looks like an erupting volcano. On cooling, the substance has the feel of coke.

❶▶ The acid is added to white sugar.

EQUATION: Dehydration of sucrose

Sucrose + concentrated sulphuric acid ➪ water as steam + carbon + sulphuric acid

$C_{12}H_{22}O_{11}(s) + H_2SO_4(conc) \Rightarrow 11H_2O(aq) + 12C(s) + H_2SO_4(aq)$

2▼ The heat given off by the reaction causes the water to turn to steam. This is trapped as bubbles by the sticky liquid.

3▼ The result is carbon that contains bubbles and looks like coke.

dehydration: the removal of water from a substance by heating it, placing it in a dry atmosphere, or through the action of a drying agent.

mineral acid: an acid that does not contain carbon and that attacks minerals. Hydrochloric, sulphuric and nitric acids are the main mineral acids.

Reaction of dilute sulphuric acid and copper

If dilute sulphuric acid (colourless) is poured on to a small pile of black copper oxide powder, the oxide reacts with the acid to produce a blue copper sulphate solution.

1▲ The acid is poured onto black copper oxide.

2◀ The copper oxide reacts and forms a blue solution of copper sulphate and water.

EQUATION: Reaction of sulphuric acid and copper oxide

Sulphuric acid + copper oxide ⇨ copper sulphate + water

$H_2SO_4(aq) + CuO(s) \Rightarrow CuSO_4(aq) + H_2O(aq)$

Nitric acid

Nitric acid (once called *aqua fortis*, or "strong water") is a colourless liquid that has been known about for many centuries. Nitric acid attacks all common metals except aluminium and iron. By mixing nitric acid and hydrochloric acid, HCl, to create a mixture known as *aqua regia*, it is possible to dissolve all metals, including gold and platinum.

Concentrated nitric acid is an oxidising chemical. Some of the oxygen it contains is readily given up to another reactant.

In the atmosphere, oxygen and nitrogen combine with the energy of a lightning flash to produce oxides of nitrogen. This, in turn, combines with rain and becomes available to green plants as dilute nitric acid, which can be used by plants as a source of nitrogen fertiliser.

Nitric acid is prepared industrially from ammonia and is used to make a wide variety of compounds, from fertilisers and dyes to explosives.

▲▶ Preparing nitric acid
This apparatus shows the preparation of fuming nitric acid from concentrated sulphuric acid and potassium nitrate.

Hot fuming nitric acid is highly corrosive. You will notice that all of the laboratory equipment being used is glass (rubber stoppers or tubing, for example would simply be corroded away).

The fuming brown gas is nitrogen dioxide.

EQUATION: Making nitric acid from ammonia

❶ *Ammonia + oxygen ⇨ nitric oxide + water*

$NH_3(g) + 5O_2(g) \Rightarrow 4NO(g) + 6H_2O(aq)$

❷ *Nitric oxide + oxygen ⇨ nitrogen dioxide*

$2NO(g) + O_2(g) \Rightarrow 2NO_2(g)$

❸ *Nitrogen dioxide + water ⇨ nitric acid + nitric oxide*

$3NO_2(g) + 6H_2O(aq) \Rightarrow 2HNO_3(aq) + NO(g)$

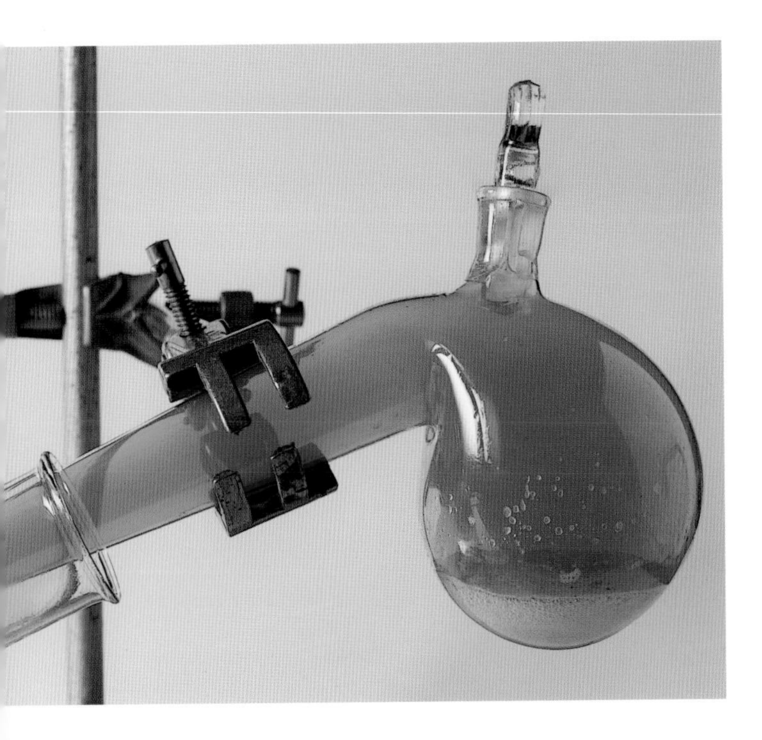

The products of nitric acid

Nitric acid reacts with a wide variety of substances. The reaction oxidises the other reagent, and reduces the nitric acid to a variety of products from nitrogen dioxide to ammonia.

Very dilute nitric acid reacts with reactive metals such as magnesium to produce hydrogen and nitric oxide. Less reactive metals such as zinc and iron produce ammonia. Poorly reactive metals such as copper and mercury result in the production of nitric oxide or nitrogen dioxide, depending on the concentration of the acid.

A few metals, such as gold and platinum, are not affected by nitric acid. Others, such as aluminium, do not react because they are protected by their oxide coating. Hydrogen sulphide is easily oxidised by nitric acid.

EQUATION: Making fuming nitric acid

Concentrated sulphuric acid + potassium nitrate ⇨ nitric acid + potassium hydrogen sulphate

$H_2SO_4(l)$ + $KNO_3(s)$ ⇨ $HNO_3(g)$ + $KHSO_4(s)$

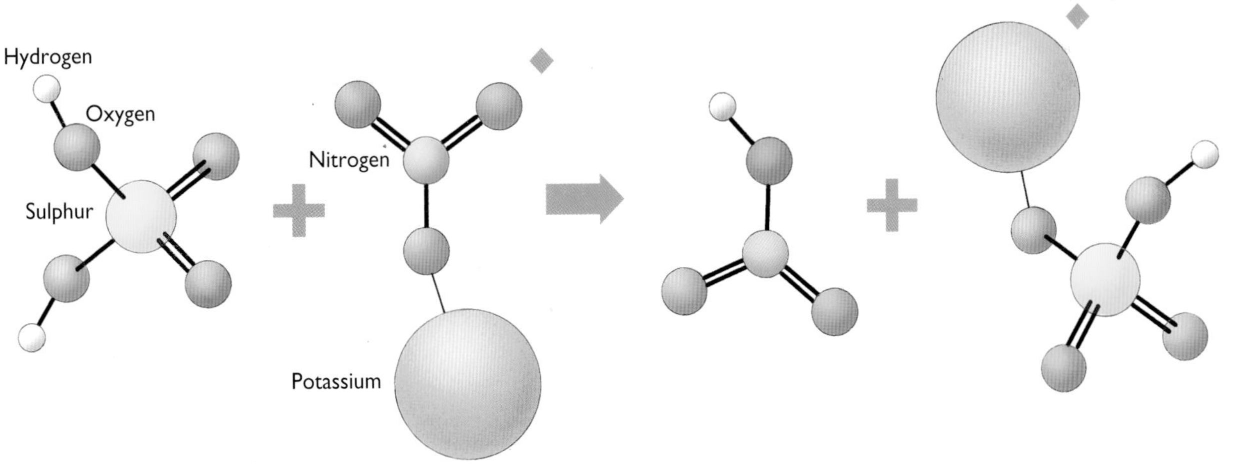

Hydrogen
Oxygen
Sulphur
Nitrogen
Potassium

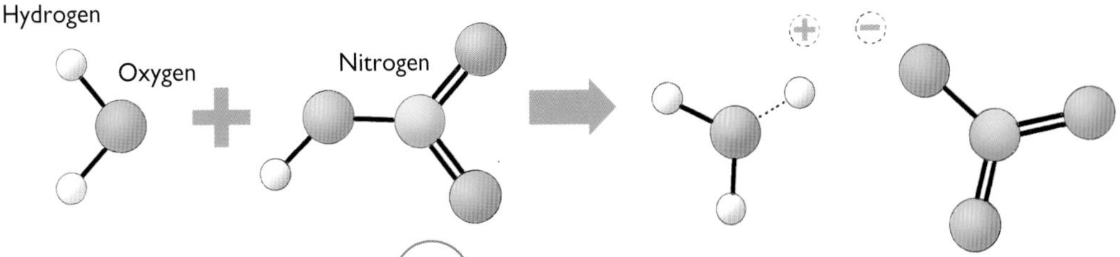

▶ Like all acids, in water nitric acid forms hydrogen ions (H^+), each of which is associated with a water molecule.

Hydrogen
Oxygen
Nitrogen

Carbonic acid

Carbonic acid, H_2CO_3, is a weak acid formed by the reaction of water and carbon dioxide gas.

Carbonic acid occurs in places where water and carbon dioxide are found naturally: in the atmosphere, where carbon dioxide gas dissolves in rainwater droplets; in the body, where it is formed in the stomach (along with hydrochloric acid), lungs and blood; and in the soil where soil organisms respire and cause a build-up of carbon dioxide gas.

Carbonic acid is responsible for the weathering of many limestone landscapes; it is partly responsible for the formation of caves and for the weathering of limestone buildings.

Carbonic acid occurs in the body as carbon dioxide reacts with water in the blood and elsewhere. As a result, the body has to have mechanisms (such as the kidneys) to regulate the amount, or an excessive build-up of carbonic acid will occur, giving rise to a disorder called acidosis.

In a healthy person, body fluids are heavily buffered with both bicarbonates and carbonic acid. The purpose of the buffers is to allow the blood to maintain itself as a neutral solution with a pH close to 7.

EQUATION: Carbonic acid

Water + carbon dioxide ⇨ carbonic acid

$H_2O(l) \quad + \quad CO_2(g) \quad ⇨ \quad H_2CO_3(aq)$

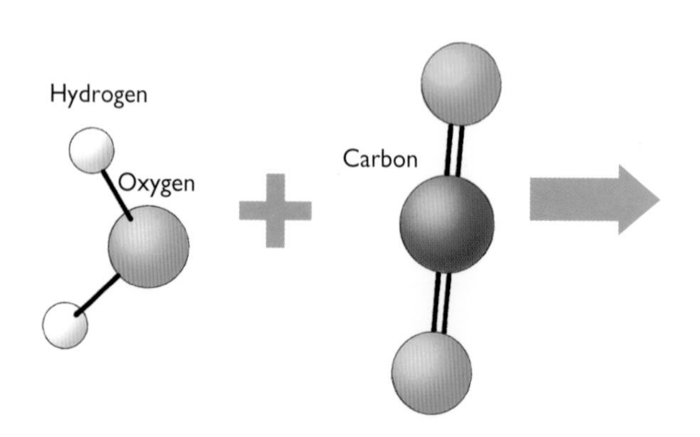

◀ The green patina, or surface coating, on copper statues, is made from copper carbonate. It is popularly called verdigris and is produced as carbonic acid and oxygen in the atmosphere react with the copper. Copper roofing displays the same effect.

buffer: a chemistry term meaning a mixture of substances in solution that resists a change in the acidity or alkalinity of the solution.

weak acid: an acid that has only partly dissociated (ionised) in water. Most organic acids are weak acids.

weather: a term used by Earth scientists and derived from "weathering", meaning to react with water and gases of the environment.

Also...

Buffer is a chemistry term meaning a mixture of substances in solution that resists a change in the acidity or alkalinity of the solution.

The purpose of a buffer is to neutralise both acids and bases. For example, the addition of a small amount of hydrochloric acid will add a large number of hydrogen ions to the solution. These will immediately react with buffer salts in the solution to produce an almost neutral result.

The reason many body fluids are heavily buffered with bicarbonates and carbonic acid is to keep body fluids neutral.

Bicarbonates

A bicarbonate is a salt derived from carbonic acid that can form as a result of a reaction between carbonic acid and a carbonate. This is shown below.

EQUATION: Dissolving limestone

Carbonic acid + calcium carbonate ➪ calcium bicarbonate

$H_2CO_3(aq)$ + $CaCO_3(s)$ ➪ $Ca(HCO_3)_2(aq)$

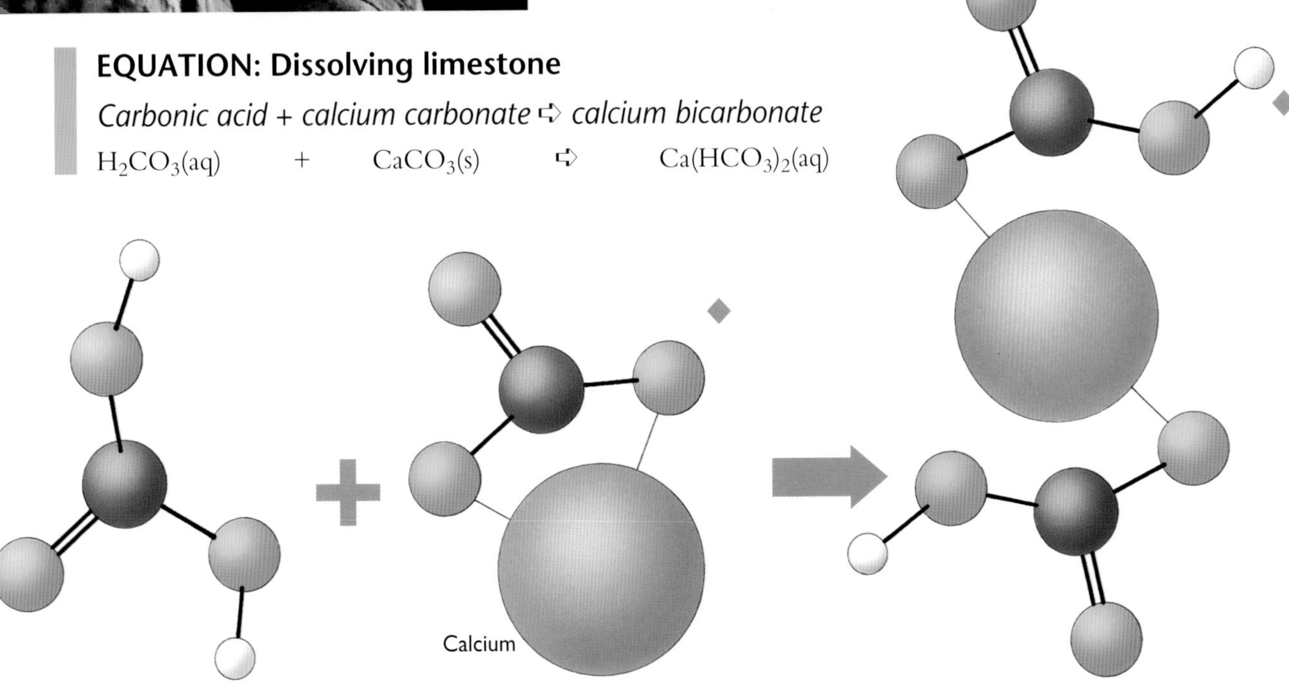

Carbonic acid

Calcium carbonate (limestone)

Calcium bicarbonate (soluble)

Organic acids

There are a large number of organic acids. Some of the most common include citric acid (the juice in citrus fruits), tartaric acid and oxalic acid.

All organic acids are weak acids, because when they are put in water only a small proportion of the acid dissociates. Organic acids therefore make poor electrolytes. It is not possible to get a concentrated solution of hydrogen ions from a weak acid.

Many organic acids (including acetic, tartaric and citric acids) fall in the group called the carboxylic acids. These acids react with strong bases such as sodium hydroxide to form salts that are more soluble in water than the acids themselves. Soap, for example, is a sodium salt that can be produced by reacting sodium hydroxide with carboxylic acid.

Fatty acids are the basis of fats and oils. Soaps, therefore, are usually manufactured from the esters in the fats and oils.

Acetic acid is important in the formation of cholesterol in the body.

Aromatic acids are another group of organic acids including, among other compounds, aspirin (a compound of salicylic acid). Ascorbic acid is vitamin C, while the bile acids act in the gut.

▼ This is a representation of tartaric acid. The two hydrogen atoms marked with the "✻" symbol will dissociate from the molecule when in water, which is what makes this compound a weak acid.

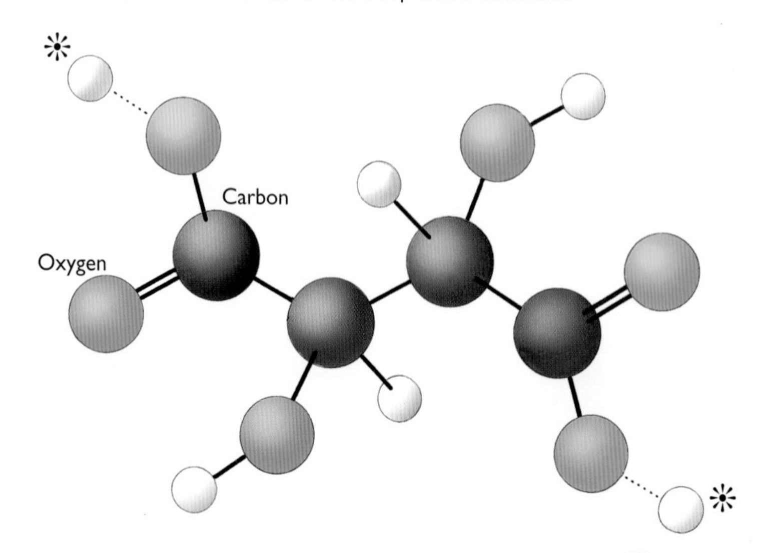

▲ Tartaric acid is found in grapes

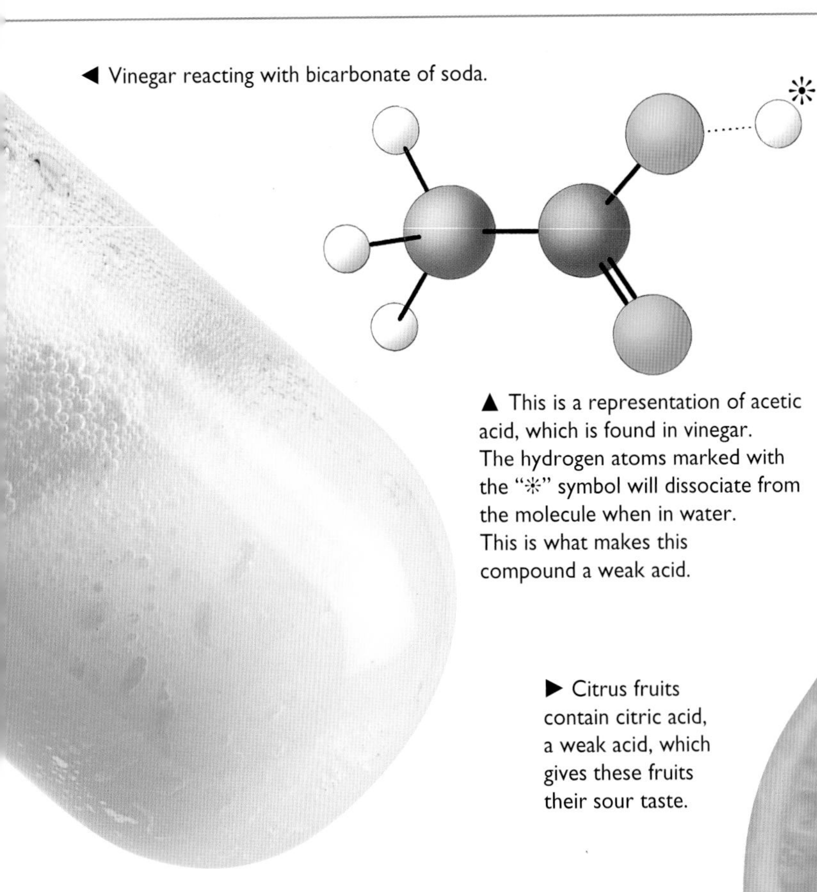

◀ Vinegar reacting with bicarbonate of soda.

▲ This is a representation of acetic acid, which is found in vinegar. The hydrogen atoms marked with the "⁎" symbol will dissociate from the molecule when in water. This is what makes this compound a weak acid.

dissociate: to break apart. In the case of acids it means to break up forming hydrogen ions. This is an example of ionisation. Strong acids dissociate completely. Weak acids are not completely ionised and a solution of a weak acid has a relatively low concentration of hydrogen ions.

electrolyte: a solution that conducts electricity.

weak acid: an acid that has only partly dissociated (ionised) in water. Most organic acids are weak acids.

▶ Citrus fruits contain citric acid, a weak acid, which gives these fruits their sour taste.

Acetic acid

Acetic acid is a colourless, weak, corrosive organic acid. Its formula is CH_3COOH. It provides the taste in vinegar.

Vinegar was first made by fermenting fruit beyond the stage needed for making wine. Wine always contains acetic acid; over-fermented wine has a distinct acid taste. Vinegar has a concentration of about 5% acetic acid.

Acetic acid reacts with various alcohols to make the range of solvents called acetates. It is also a raw material of the artificial fibre, rayon. Acetylsalicylic acid is better known as the painkiller aspirin.

Industrial acetic acid is made from carbon monoxide and hydrogen. The first step is to form methyl alcohol from hydrogen and carbon monoxide. Adding excess carbon monoxide creates acetic acid.

Citric acid

Citric acid is a solid, white, organic acid. It is found in all animal and plant cells where it is used as part of the process of metabolism. It is especially rich in citrus fruits, especially lemons and limes. It is more acidic (i.e. it has a greater concentration of H^+ ions in solution) than acetic acid.

The formula for citric acid is HOOCCH 2C(OH)(COOH)CH 2COOH H_2O. Citric acid can be made by decomposing molasses from sugar beet using mould. The acid occurs as part of natural fermentation.

Citric acid is used industrially as a preservative in food, as a cleaning and polishing agent for metal and in soft drinks and flavourings.

Bases

Bases make up a large and important group of chemicals. If a base is water-soluble it gives an alkaline reaction when tested with a chemical indicator. Bases are metal oxides and hydroxides, and nearly all are insoluble. Bases contain hydroxide ions, an ion consisting of one hydrogen and one oxygen atom.

If a base is soluble, like sodium hydroxide (caustic soda), it is called an alkali. An alkali has an excess of hydroxide ions (the opposite to an acid, which has an excess of hydrogen ions). Like an acid, a strong alkali will completely dissociate in water, releasing the ions of its metal and its hydroxide.

In general, a base will react with an acid to form a new substance (called a salt) and water but no additional compounds. Sodium chloride (common salt) is a neutral substance that will show neither acid nor alkaline response when tested with a chemical indicator.

Examples of bases are ammonia (NH_3, a gas in solution in water used as a cleaning agent and in fertilisers), lime (CaO, a solid used in cement and fertilisers), magnesium hydroxide ($Mg(OH)_2$, a solid used as an antacid for indigestion) and sodium hydroxide ($NaOH$, a solid used in oven cleaners and soap-making). Some bases, such as sodium hydroxide, are caustic and harmful to the skin.

Most bases are insoluble
The bases formed in the test tubes shown here are all insoluble in water, making coloured precipitates.

Each sample was prepared by adding sodium hydroxide to solutions of the sulphates of the metals cobalt, copper and iron.

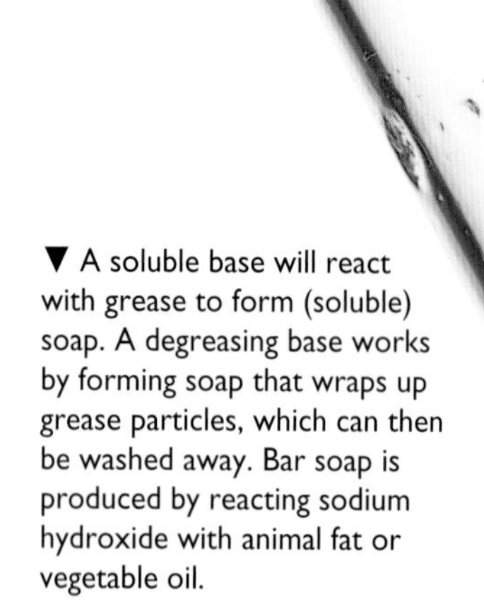

▼ A soluble base will react with grease to form (soluble) soap. A degreasing base works by forming soap that wraps up grease particles, which can then be washed away. Bar soap is produced by reacting sodium hydroxide with animal fat or vegetable oil.

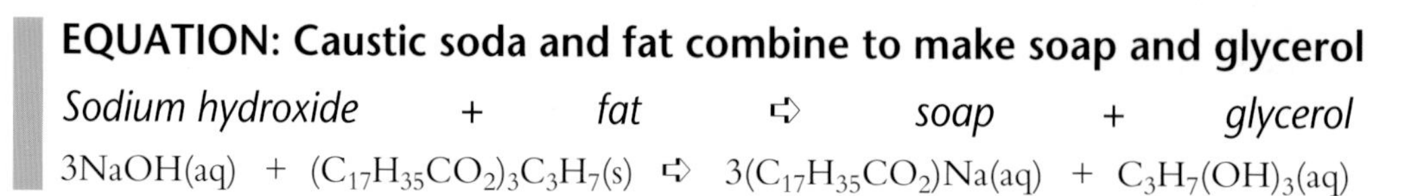

EQUATION: Caustic soda and fat combine to make soap and glycerol

Sodium hydroxide + *fat* ⇨ *soap* + *glycerol*

$3NaOH(aq) + (C_{17}H_{35}CO_2)_3C_3H_7(s) \Rightarrow 3(C_{17}H_{35}CO_2)Na(aq) + C_3H_7(OH)_3(aq)$

alkaline: the opposite of acidic. Alkalis are bases that dissolve, and alkaline materials are called basic materials. Solutions of alkalis have a pH greater than 7.0 because they contain relatively few hydrogen ions.

base: a compound that may be soapy to the touch and that can react with an acid in water to form a salt and water.

salts: compounds, often involving a metal, that are the reaction products of acids and bases. (Note "salt" is also the common word for sodium chloride, common salt or table salt. It is therefore important to read the context of the word carefully to get the correct meaning.)

The reaction of acids with bases

The reaction of acids and bases is very common. For example when naturally acidic rainwater falls on to (basic) limestone (calcium carbonate), the products are a salt, water and carbon dioxide. We know this reaction as weathering. It is the action that changes solid rocks to small particles and soluble substances that can readily be moved by running water. Such reactions have created the landscape all around us. The same thing happens when acid rain falls on the land, although in this case, neutralising the acid may be at the expense of nutrients in the soil.

In many industrial processes, chemists react one substance with another to produce useful products. It is also important to use just the right amount of each reagent so that nothing is wasted. In some cases it is also vital that the product be neutral (shampoos and face creams for example).

EQUATION: Titration of hydrochloric acid and sodium hydroxide

Sodium hydroxide + hydrochloric acid ➪ sodium chloride + water

$NaOH(aq)$ + $HCl(aq)$ ➪ $NaCl(aq)$ + $H_2O(l)$

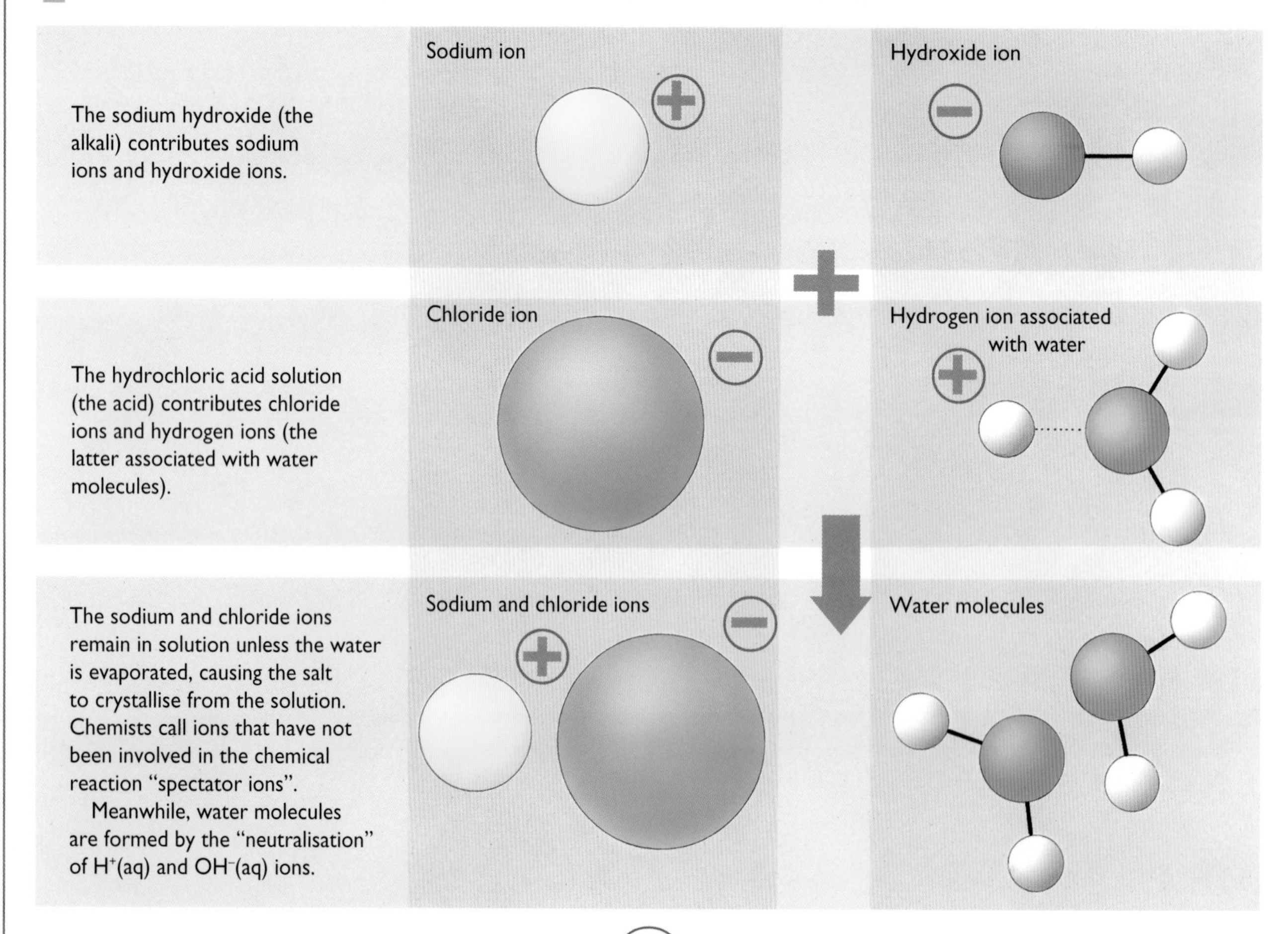

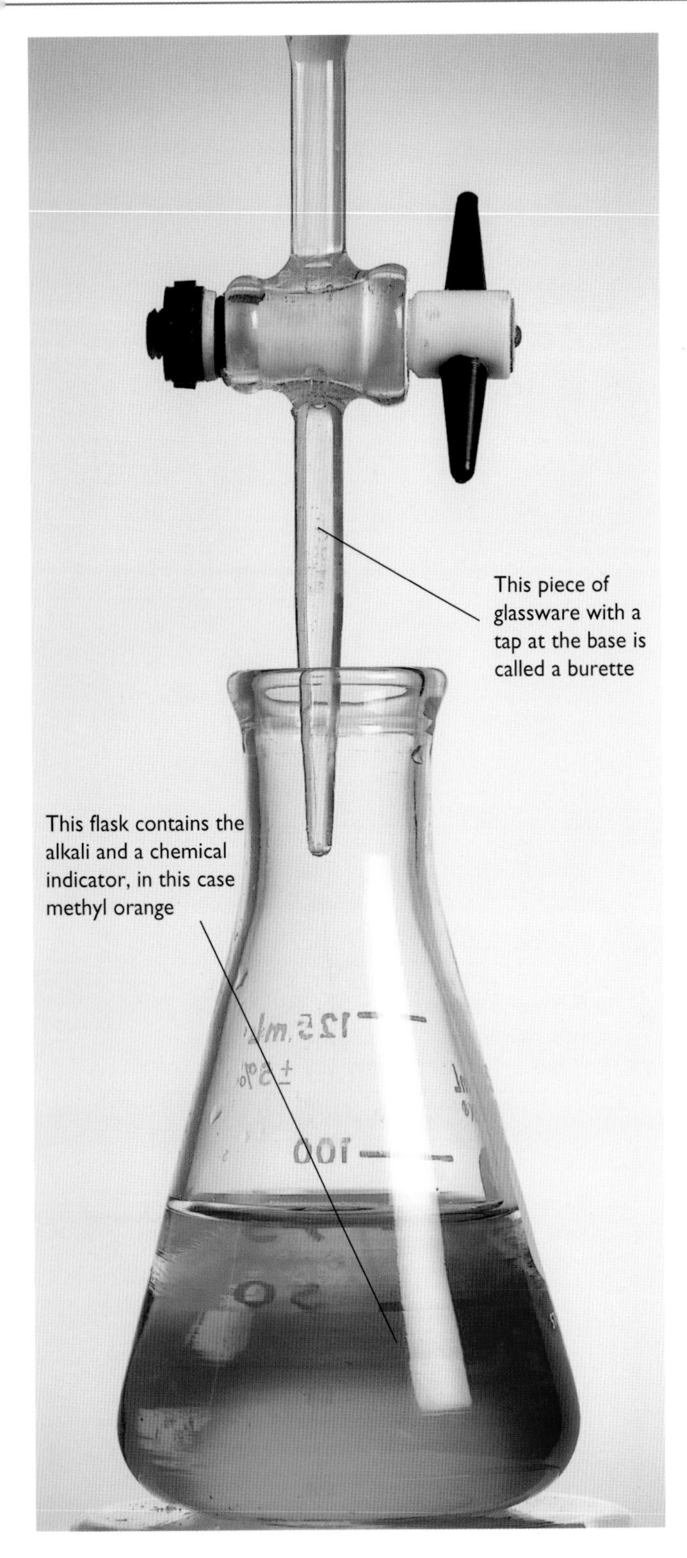

ion: an atom, or group of atoms, that has gained or lost one or more electrons and so developed an electrical charge.

titration: a process of dripping one liquid into another in order to find out the amount needed to cause a neutral solution. An indicator is used to signal change.

Neutralising

When an acid and an alkali react, the hydrogen ions of the acid react with the hydroxide ions of the alkali to produce water. If there are appropriate amounts of acid and alkali to react, then the result will be a neutral substance (a salt) and water. For example, when equal amounts of sodium hydroxide (NaOH) and hydrochloric acid (HCl) are mixed, the hydrogen ions and hydroxide ions react to form water, leaving only the sodium ions and chloride ions in solution. The result is a neutral salt solution (sodium chloride).

If the amount of acid exceeds the alkali then the reaction will still occur, but after all reacting has been completed, the result will still be an acidic solution. Similarly, if the amount of acid is less than the alkali, then the result will be an alkaline solution. When scientists want to find out how much of one chemical is needed to neutralise another, they use a method called titration, which is shown here.

Also...
How to find out the amount of acid needed to neutralise a base

During the process of adding acid to alkali (a process chemists call titration) an indicator is used to show when the acid introduced from a burette exactly neutralises the alkali in a flask. At this point the hydrogen ions from the acid in the flask exactly match the number of hydroxide ions from the alkali and the liquid turns from yellow to pink.

In the example shown here the solution has not been shaken so that you can see both the pink and yellow colours of the indicator. It is, however, normal practice to shake the flask continually so that the indicator shows a change throughout the liquid.

EQUATION: Titration of hydrochloric acid and sodium hydroxide

Hydrochloric acid + sodium hydroxide ⇨ sodium chloride + water

$HCl(aq) + NaOH(aq) \Rightarrow NaCl(aq) + H_2O(l)$

The reaction of acids with metals

The reaction of acids with metals is often referred to as corrosion. It is an extremely important reaction. It is also a very good demonstration of the use of the Periodic Table (see page 46).

Metals are on the left of the Periodic Table. They have one, two or three more electrons than the nearest inert gas. When metals react (corrode) they lose these outer shell electrons and become positively charged ions.

Each electron lost means one more positive charge on the ion. For example, sodium (Na^{+}) has lost one electron, magnesium (Mg^{2+}) has lost two and aluminium (Al^{3+}) three.

For these metals to react, the "lost" electrons have to be accepted by some other substance. Hydrogen ions (H^{+}, produced by acids in water) can readily accept electrons, turning hydrogen ions into hydrogen molecules.

Each hydrogen ion gains one electron to become a hydrogen atom and then two hydrogen atoms share their electrons to become a hydrogen molecule and thus form hydrogen gas.

Some examples of reactions between metals and acids are shown on these two pages.

▲▶▼ For a metal to react in such a way as to release hydrogen gas the metal has to be above hydrogen in the reactivity series.

REACTIVITY SERIES	
Element	*Reactivity*
potassium	*most reactive*
sodium	
calcium	
magnesium	
aluminium	
manganese	
chromium	
zinc	
iron	
cadmium	
tin	
lead	
hydrogen	
copper	
mercury	
silver	
gold	
platinum	*least reactive*

EQUATION: Zinc and sulphuric acid

Zinc + sulphuric acid ➪ zinc sulphate + water + hydrogen gas

$Zn(s) + H_2SO_4(aq) \Rightarrow ZnSO_4(s) + H_2O(l) + H_2(g)$

Hydrogen gas

►▲ Metals below hydrogen in the reactivity series, such as mercury, only react with an acid that is also an oxidising agent. The product is then not hydrogen gas but water.

Brown nitrogen dioxide gas

Concentrated nitric acid is poured onto mercury

EQUATION: Mercury and nitric acid

Mercury + concentrated nitric acid ➪ mercury nitrate + water + nitrogen dioxide

$Hg(s) + 4HNO_3(conc) \Rightarrow Hg(NO_3)_2(aq) + 2H_2O(l) + 2NO_2(g)$

The noble gases

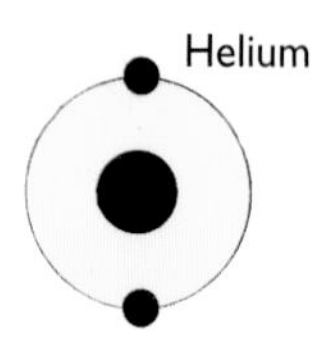

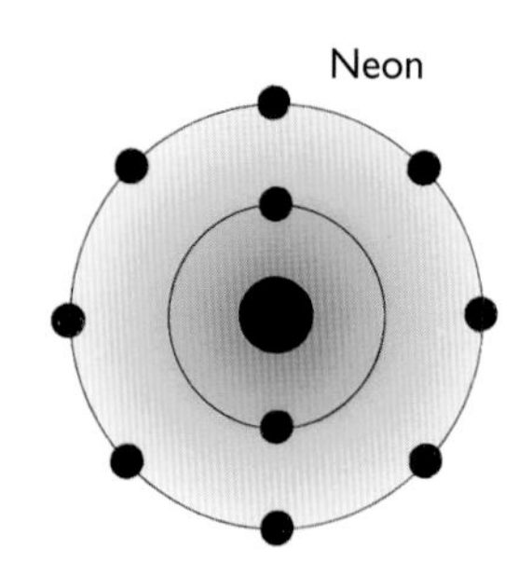

The noble, or inert, gases are helium, neon, argon, krypton, xenon and radon. They make up a complete group (0) in the Periodic Table and the properties of each member of the group are closely related to the others. The noble gases make up nine-tenths of 1% of the volume of the Earth's atmosphere.

The noble gases are called inert gases because they do not usually react with any other element. Most elements react with one another because the combined atoms are more stable than the individual atoms. In fact, the most stable state generally occurs when the number of electrons in the outermost shell of an atom is eight. Reactions occur so that atoms can achieve eight electrons, either by losing electrons or gaining them from their partner. The noble gases are so stable, and react so little because, unusually for elements, the number of electrons in the outer part of the atom completely fill the shell. In all but helium the number of electrons in the outer shell is eight; for helium it is two – the maximum possible in its only shell.

Not all of the noble gases are equally stable. For example, xenon can be made to combine with a few other elements, but helium combines with none.

Sir William Ramsay and the noble gases

Scientists can now tell what elements make up a substance by looking at the light it gives out when heated. The characteristic pattern of light is called a spectrum. Each element has a unique spectrum that can be readily picked out by an instrument called a spectrometer.

In the middle of the last century when scientists looked at the wavelengths of light being given out by the Sun, they could identify all of them except for a mysterious yellow light. This was helium. Helium was discovered on Earth in a sample of rock in 1895 by a Scotsman, William Ramsay, who then went on to discover all the other noble gases.

▼ Sir William Ramsay experimenting with radium in his laboratory.

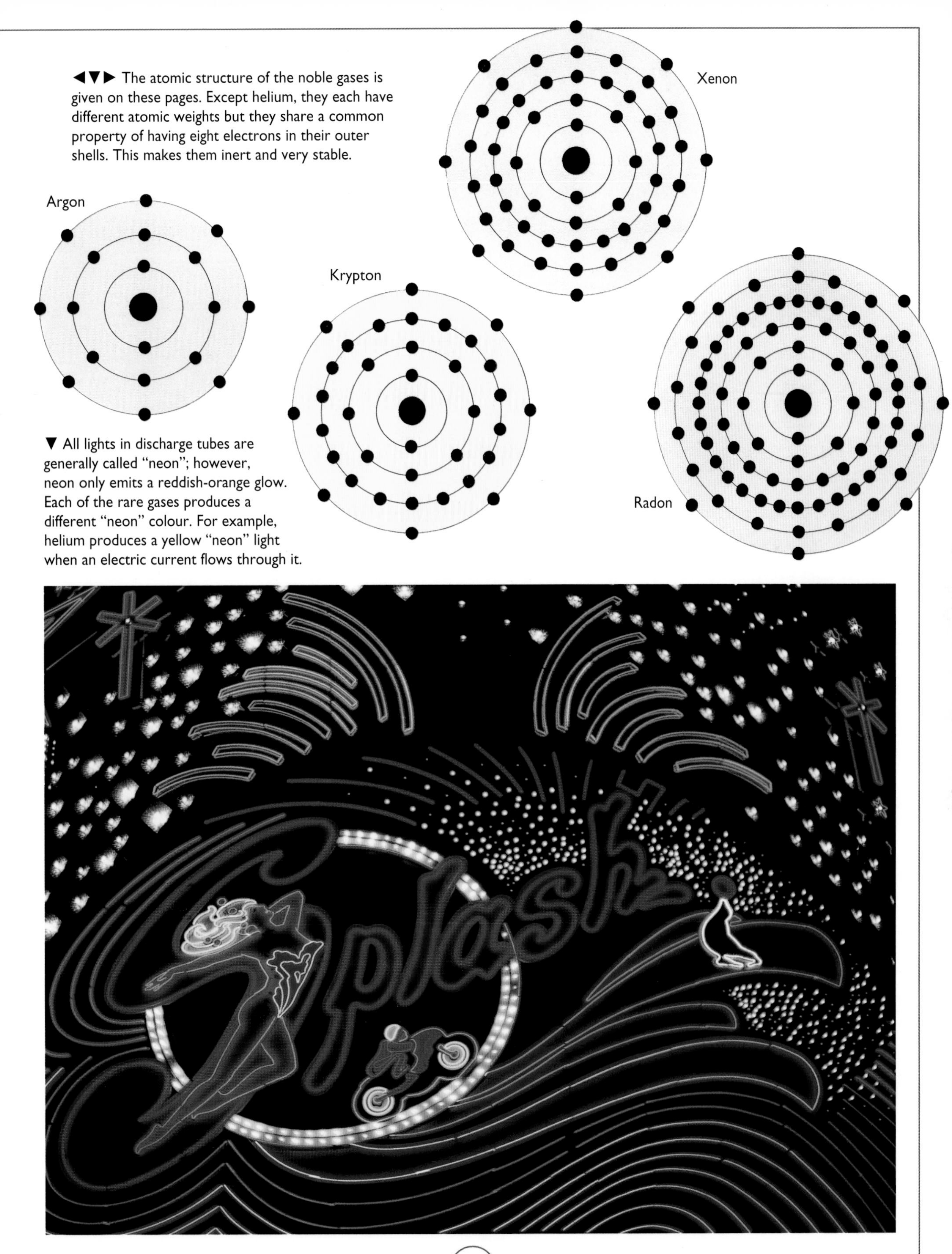

◀▼▶ The atomic structure of the noble gases is given on these pages. Except helium, they each have different atomic weights but they share a common property of having eight electrons in their outer shells. This makes them inert and very stable.

▼ All lights in discharge tubes are generally called "neon"; however, neon only emits a reddish-orange glow. Each of the rare gases produces a different "neon" colour. For example, helium produces a yellow "neon" light when an electric current flows through it.

Helium and neon

Helium is named after the Greek word *helios*, meaning "Sun", so named because the gas was first observed in the Sun.

It is one of the smallest and lightest of the elements and the first of the noble gases. Because it is chemically inert, it does not form compounds and so does not get locked away in rocks. Being so light, it soon floats upwards through the air and escapes to space.

Neon, symbol Ne, is the second of the noble gases (after helium) in the Periodic Table and was named for the Greek *neos* meaning "new". It makes up about one part in 65,000 parts of the Earth's atmosphere.

Where does helium come from?

Helium is the stuff of stars, and was formed, some scientists believe, in the first few minutes of the creation of the Universe (known as the Big Bang). It is still being made as hydrogen fuses together to fuel the energy released from the stars. Helium is also being made on Earth. This happens because some radioactive substances decay to release helium, and it is this tiny amount of product that becomes occasionally trapped, along with natural gas, in reservoirs below the ground. It can therefore be recovered along with some natural gas deposits.

Uses for neon

Until recently, liquid helium has been thought to be the best gas to use for refrigeration systems, but neon is over 40 times as effective, so this application may increase in the future.

Neon is best known for giving its name to the glowing discharge tubes called "neon tubes". In fact neon glows with a reddish–orange colour, and the other colours seen in neon lights are in fact discharges from other noble gases.

Neon is also used in many other applications, including television tubes and gas lasers.

▶ Helium is used to fill party balloons and modern airships because it is safer. It has almost the same lifting power as hydrogen, but cannot catch fire. Of all the helium produced 10% is used for lifting.

Collecting helium

We cannot make helium by any chemical means because it is not locked in any compounds, so all the helium we use has to be collected, along with natural gas, from gas fields. The majority of the world's helium comes from Texas, Utah, Oklahoma and Kansas.

inert: nonreactive. There are few substances that hardly react. They include all the rare gases (helium, neon, argon, krypton, xenon and radon).

▲ The extremely low boiling point of helium makes it useful for studying matter at very low temperatures, a type of research known as cryogenics (from the Greek word for "frost").

Cold helium

Helium boils below any other element. It can only be liquified at -269°C, just four degrees above absolute zero, the coldest it is possible to get. Helium is also the one element that cannot be made to go into a solid at normal pressures. It remains liquid even close to absolute zero.

At these low temperatures helium behaves very strangely, having an ability to conduct heat 600 times as well as copper, flowing uphill, climbing up and over the walls of the flask containing it, moving almost without friction and going through the tiniest of holes. For these reasons, some scientists think that cold helium may be a fourth state of matter.

Helium is the best refrigerant in the Universe. It is used on space shuttles to cool the hydrogen and oxygen fuels, both of which have to be kept in liquid form during the journey. So helium, the gas of space, has a vital role to play in affecting our first steps into space.

The main use of helium on Earth is for cooling down scientific equipment for special purposes. Body scanners, for example, rely on helium.

Airship

An airship is a type of lighter-than-air craft using a balloon for buoyancy and an engine to propel it. It carries a small passenger compartment – known as a gondola – below the balloon. Airships used to be filled with hydrogen because of the excellent lift this provided, but the fire hazards with hydrogen mean that all modern balloons are filled with helium.

Airships have to float at a desired height. Old airships used ballast that could be thrown overboard to allow the airship to rise. To make it sink, hydrogen gas was released from the balloon.

Helium is too precious to be released in this way, and throwing out ballast is hazardous for those below. Extra ballast is now provided by causing water vapour in the air to condense, adding weight by increasing the amount of water condensed.

Also...

A mixture of 80% helium and 20% oxygen is used as an artificial atmosphere for deep sea diving. Helium is used instead of nitrogen because it does not cause the "bends". Its use has a remarkable effect on the vocal chords, making people who breathe the gas sound like Mickey Mouse!

Argon, krypton, xenon and radon

Argon is the third of the noble gases in the periodic table. It is much more abundant than the other noble gases, making up the great majority of the noble gases in the atmosphere. It is 0.94% of the air by volume (the total volume of all the noble gases is less than 1%).

Like all the noble gases, argon is obtained for industrial use by cooling air down to very low temperatures, and then collecting the element when it changes from gas to liquid at -185.7°C.

The fourth, fifth and sixth members of the noble gases are krypton, xenon and radon. They have higher atomic numbers than the other members of the group, more compounds associated with them, and radioactive forms (called isotopes). Radon is composed entirely of radioactive isotopes. Some of the radioactive isotopes of krypton are made as part of the nuclear processes that occur during electric power generation. Radon is colourless, but xenon gives blue, and krypton violet, light when excited by an electrical current.

Xenon

Xenon, symbol Xe, which is a Greek word meaning "a stranger", is the fifth member of the group of noble gases in the Periodic Table. It is the rarest of all the gases, being present in the atmosphere at a concentration of only one part in twenty million by volume.

It produces a bright blue glow when used as the gas in a discharge tube. The light emitted from high-intensity bulbs is a bluish-white colour and is widely used as the source of light for lighthouses and for stroboscopes used at rock concerts.

Uses for argon

Argon is used together with nitrogen in incandescent tubes and also in fluorescent tubes. It is also used in factories making silicon wafers, providing an inert environment in which the silicon crystals can grow as they are slowly extracted from the molten silicon.

▶ A mixture of argon and nitrogen is used in incandescent light bulbs. When the filament gets hot, it sends out atoms into the bulb. If the bulb contained a reactive gas the life of the filament would be reduced, and a thin film of metal atoms would be deposited on the inside of the bulb, blackening it and reducing the efficiency of the light.

Argon and nitrogen are inert gases and so do not change in the intense heat of the filament. The presence of gas molecules also means that the atoms leaving the filament are more likely to encounter the gas and bounce back on the filament than they are to reach the glass of the bulb. This prolongs the life of the bulb and stops it from blackening.

▶ The intense light source from a lighthouse is provided by a discharge tube filled with xenon.

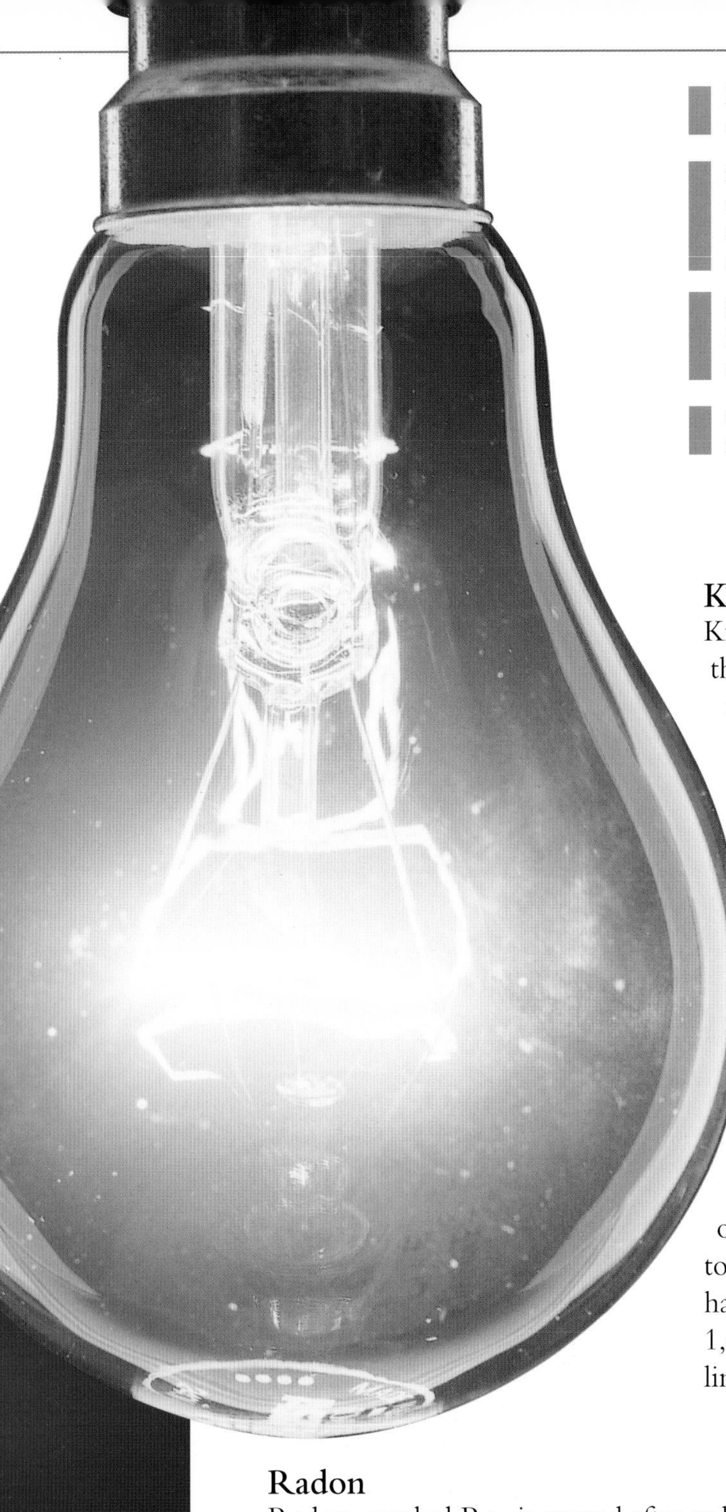

half-life: the time it takes for the radiation coming from a sample of a radioactive element to decrease by half.

inert: nonreactive. There are few substances that hardly react. They include all the rare gases (helium, neon, argon, krypton, xenon and radon) and also diamonds and quartz.

isotope: atoms that have the same number of protons in their nucleus, but which have different masses; for example, carbon-12 and carbon-14.

radioactive: a material that emits radiation or particles from the nucleus of its atoms.

Krypton

Krypton, symbol Kr, whose name comes from the Greek *kryptos*, meaning "hidden", is the fourth of the noble gases in the periodic table. It occurs as only one part in every million parts of air by volume.

As with other rare gases, for industrial purposes krypton is extracted by cooling air until the gas turns to liquid. In the case of krypton this is -152°C. However, it is very difficult to isolate krypton this way, so the gas is very expensive. This factor, more than any other, limits the number of uses to which krypton is put. It is used in both discharge tubes and incandescent bulbs.

Krypton also has the unusual distinction of being the element whose light is used to define the metre. Since 1960, scientists have defined the metre to be the distance of 1,650,763.73 wavelengths of the orange–red line of krypton, atomic weight 86.

Radon

Radon, symbol Rn, is named after radium. It is the sixth member of the noble gases in the Periodic Table. Radon is very different from the other rare gases insofar as it is radioactive, produced by the metal element radium. Radon gas is common worldwide, although it is more concentrated in areas where there are igneous rocks such as granites. The surface soils of the world contain about one gram of radium-emitting radon in every three square kilometres.

The half-life of radon is just under four days. The main form of radiation is alpha particles, the least harmful and least penetrating form of radiation.

Key facts about...

Hydrogen

A colourless gas, chemical symbol H

Liquid hydrogen is one-tenth as dense as water

Low density gas (less than one-twelfth as dense as air)

Can be explosive and inflammable

Has no taste

Has no smell

Its isotope tritium is radioactive

Insoluble in water

Its isotope deuterium is radioactive and known as "heavy water"

Comprises about 99.88% of the Universe

Atomic number 1, atomic weight 1

SHELL DIAGRAMS

The shell diagrams on these pages are representations of an atom of each element. The total number of electrons are shown in the relevant orbitals, or shells, around the central nucleus.

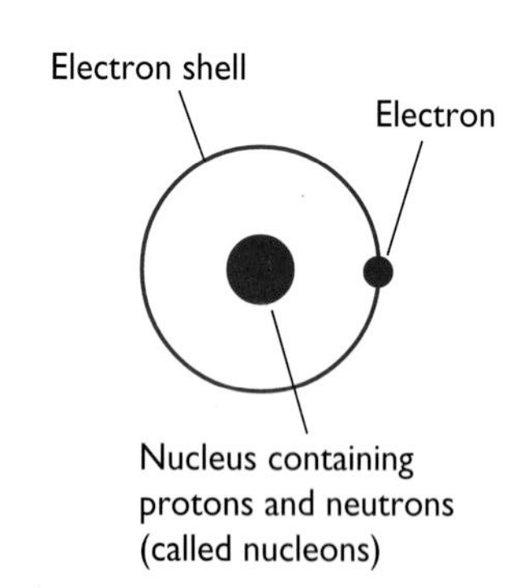

Helium

A colourless gas, chemical symbol He

Has no smell

Inert

Has no taste

Nontoxic

Atomic number 2, atomic weight about 4

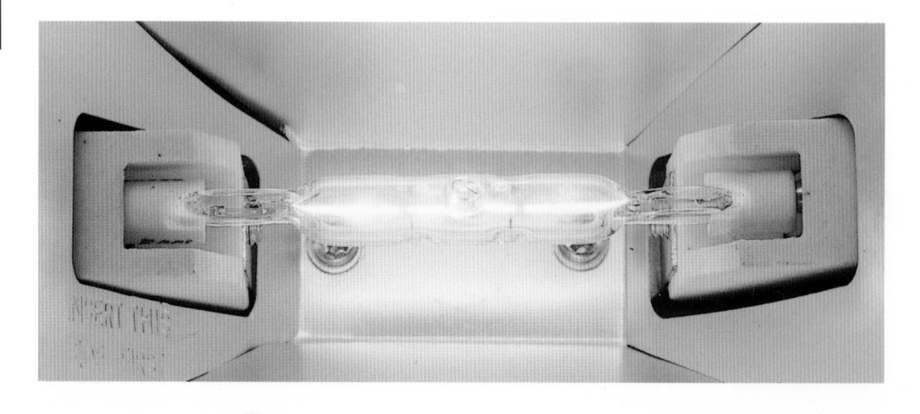

► Halogen lamps are filled with a mixture of xenon and a halogen vapour. They are essentially incandescent lights.

Neon

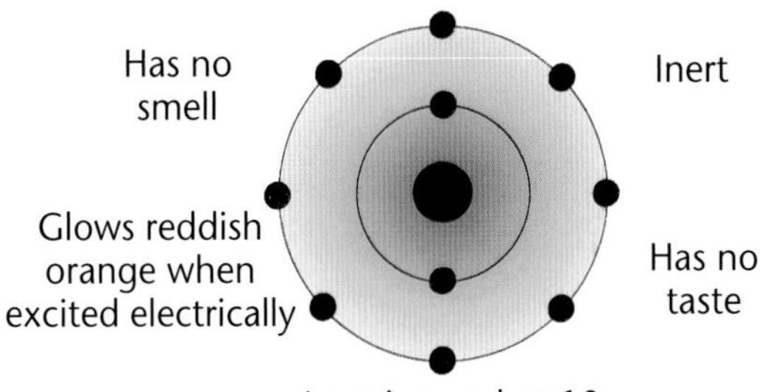

Argon

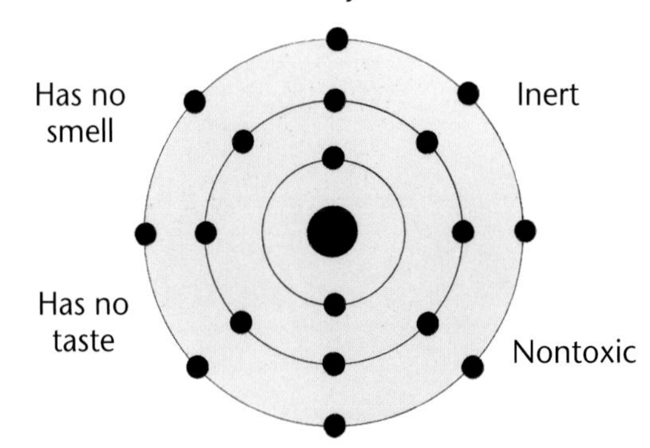

Krypton

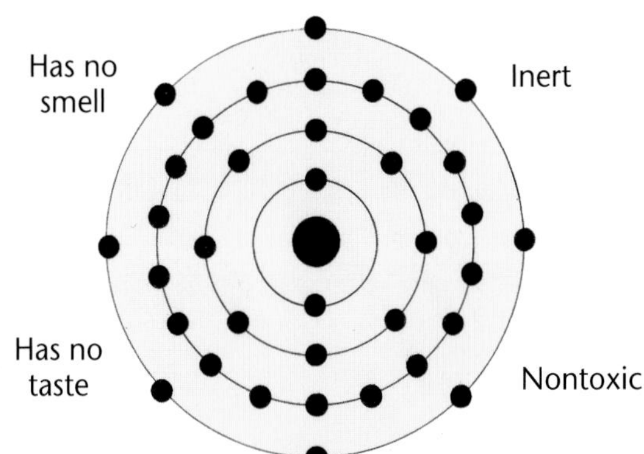

Xenon

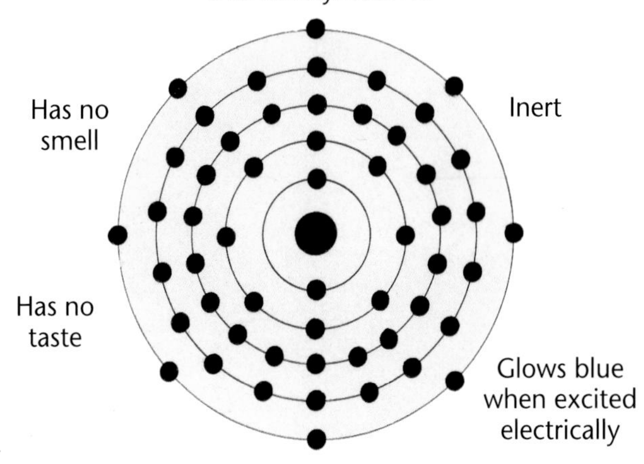

Radon

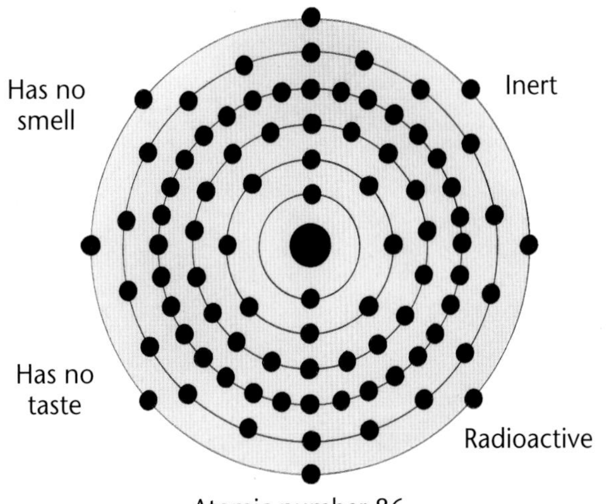

The Periodic Table

The Periodic Table sets out the relationships among the elements of the Universe. According to the Periodic Table, certain elements fall into groups. The pattern of these groups has, in the past, allowed scientists to predict elements that had not at that time been discovered. It can still be used today to predict the properties of unfamiliar elements.

The Periodic Table was first described by a Russian teacher, Dmitry Ivanovich Mendeleev, between 1869 and 1870. He was interested in writing a chemistry textbook, and wanted to show his students that there were certain patterns in the elements that had been discovered. So he set out the elements (of which there were 57 at the time) according to their known properties. On the assumption that there was pattern to the elements, he left blank spaces where elements seemed to be missing. Using this first version of the Periodic Table, he was able to predict in detail the chemical and physical properties of elements that had not yet been discovered. Other scientists began to look for the missing elements, and they soon found them.

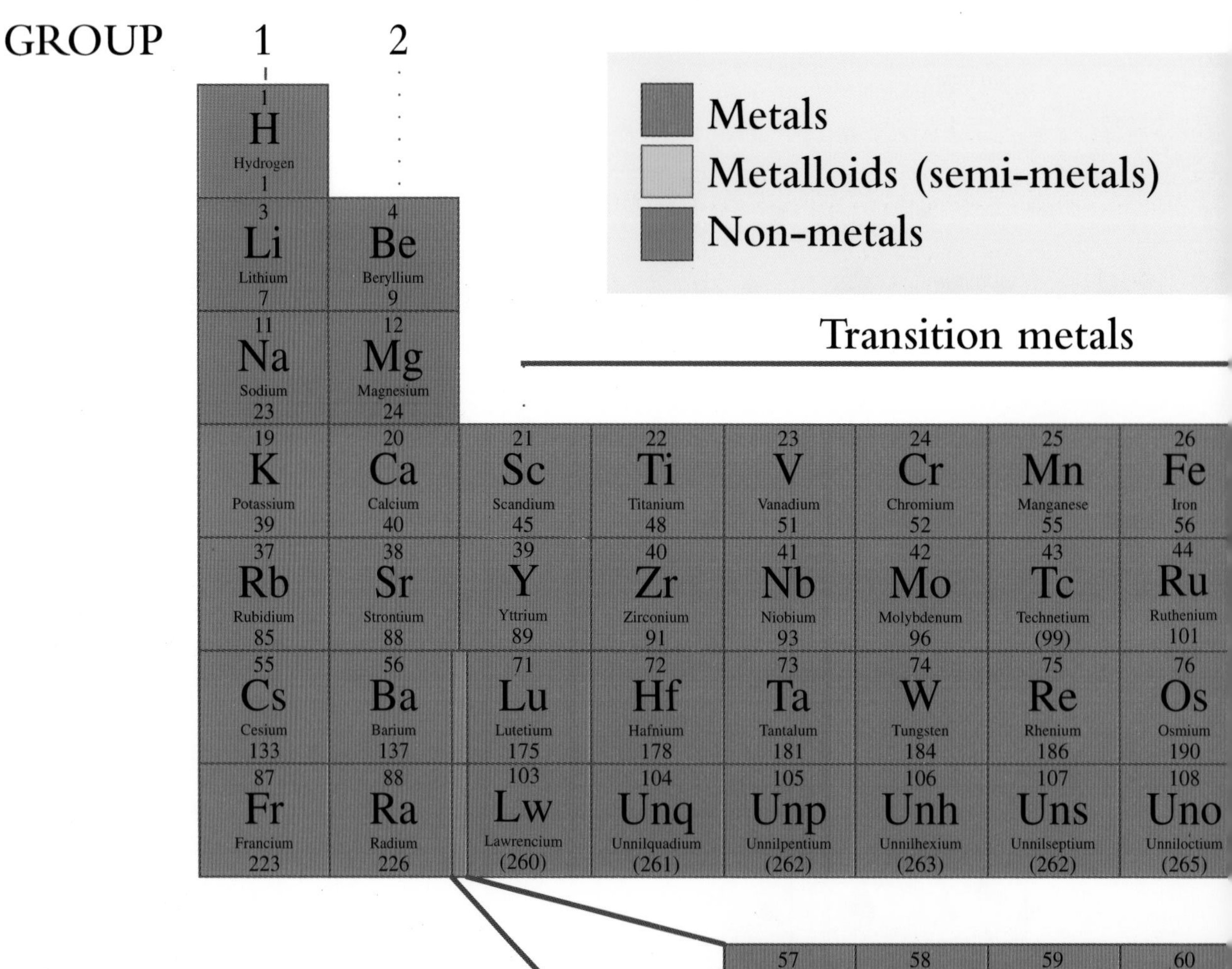

Hydrogen did not seem to fit into the table, so he placed it in a box on its own. Otherwise the elements were all placed horizontally. When an element was reached with properties similar to the first one in the top row, a second row was started. By following this rule, similarities among the elements can be found by reading up and down. By reading across the rows, the elements progressively increase their atomic number. This number indicates the number of positively charged particles (protons) in the nucleus of each atom. This is also the number of negatively charged particles (electrons) in the atom.

The chemical properties of an element depend on the number of electrons in the outermost shell.

Atoms can form compounds by sharing electrons in their outermost shells. This explains why atoms with a full set of electrons (like helium, an inert gas) are unreactive, whereas atoms with an incomplete electron shell (such as chlorine) are very reactive. Elements can also combine by the complete transfer of electrons from metals to non-metals and the compounds formed contain ions.

Radioactive elements lose particles from their nucleus and electrons from their surrounding shells. As a result their atomic number changes and they become new elements.

Atomic (proton) number
13 Al Aluminium 27
Symbol
Name
Approximate relative atomic mass
(Approximate atomic weight)

				3	4	5	6	7	0
									2 He Helium 4
				5 B Boron 11	6 C Carbon 12	7 N Nitrogen 14	8 O Oxygen 16	9 F Fluorine 19	10 Ne Neon 20
				13 Al Aluminium 27	14 Si Silicon 28	15 P Phosphorus 31	16 S Sulphur 32	17 Cl Chlorine 35	18 Ar Argon 40
27 Co Cobalt 59	28 Ni Nickel 59	29 Cu Copper 64	30 Zn Zinc 65	31 Ga Gallium 70	32 Ge Germanium 73	33 As Arsenic 75	34 Se Selenium 79	35 Br Bromine 80	36 Kr Krypton 84
45 Rh Rhodium 103	46 Pd Palladium 106	47 Ag Silver 108	48 Cd Cadmium 112	49 In Indium 115	50 Sn Tin 119	51 Sb Antimony 122	52 Te Tellurium 128	53 I Iodine 127	54 Xe Xenon 131
77 Ir Iridium 192	78 Pt Platinum 195	79 Au Gold 197	80 Hg Mercury 201	81 Tl Thallium 204	82 Pb Lead 207	83 Bi Bismuth 209	84 Po Polonium (209)	85 At Astatine (210)	86 Rn Radon (222)
109 Une Unnilennium (266)									

61 Pm Promethium (145)	62 Sm Samarium 150	63 Eu Europium 152	64 Gd Gadolinium 157	65 Tb Terbium 159	66 Dy Dysprosium 163	67 Ho Holmium 165	68 Er Erbium 167	69 Tm Thulium 169	70 Yb Ytterbium 173
93 Np Neptunium (237)	94 Pu Plutonium (244)	95 Am Americium (243)	96 Cm Curium (247)	97 Bk Berkelium (247)	98 Cf Californium (251)	99 Es Einsteinium (252)	100 Fm Fermium (257)	101 Md Mendelevium (258)	102 No Nobelium (259)

Understanding equations

As you read through this book, you will notice that many pages contain equations using symbols. If you are not familiar with these symbols, read this page. Symbols make it easy for chemists to write out the reactions that are occurring in a way that allows a better understanding of the processes involved.

Symbols for the elements

The basis of the modern use of symbols for elements dates back to the 19th century. At this time a shorthand was developed using the first letter of the element wherever possible. Thus "O" stands for oxygen, "H" stands for hydrogen and so on. However, if we were to use only the first letter, then there could be some confusion. For example, nitrogen and nickel would both use the symbols N. To overcome this problem, many elements are symbolised using the first two letters of their full name, and the second letter is lowercase. Thus although nitrogen is N, nickel becomes Ni. Not all symbols come from the English name; many use the Latin name instead. This is why, for example, gold is not G but Au (for the Latin *aurum*) and sodium has the symbol Na, from the Latin *natrium*.

Compounds of elements are made by combining letters. Thus the molecule carbon

Written and symbolic equations

In this book, important chemical equations are briefly stated in words (these are called word equations), and are then shown in their symbolic form along with the states.

What reaction the equation illustrates

EQUATION: The formation of calcium hydroxide

Word equation — *Calcium oxide + water ⇨ calcium hydroxide*

Symbol equation — $CaO(s) + H_2O(l) \Rightarrow Ca(OH)_2(aq)$
heated

Sometimes you will find additional descriptions below the symbolic equation.

Symbol showing the state: *s* is for solid, *l* is for liquid, *g* is for gas and *aq* is for aqueous.

Diagrams

Some of the equations are shown as graphic representations.

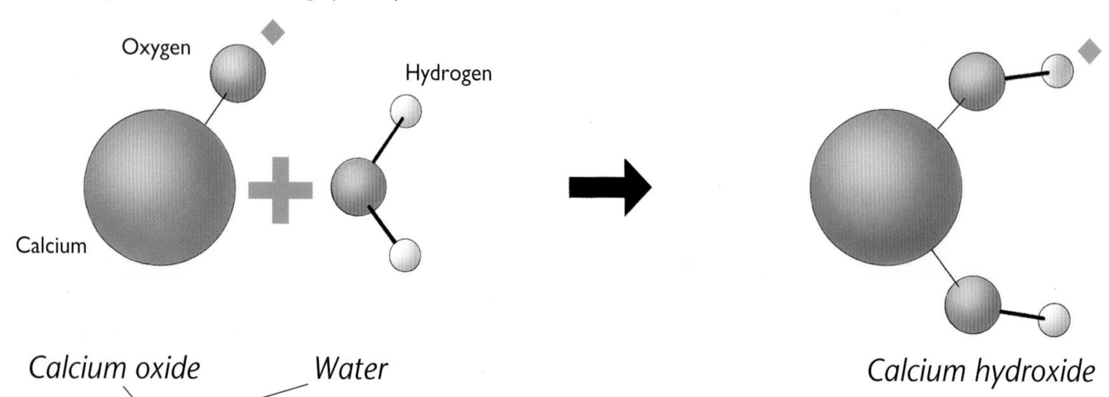

Sometimes the written equation is broken up and put below the relevant stages in the graphic representation.

monoxide is CO. By using lowercase letters for the second letter of an element, it is possible to show that cobalt, symbol Co, is not the same as the molecule carbon monoxide, CO.

However, the letters can be made to do much more than this. In many molecules, atoms combine in unequal numbers. So, for example, carbon dioxide has one atom of carbon for every two of oxygen. This is shown by using the number 2 beside the oxygen, and the symbol becomes CO_2.

In practice, some groups of atoms combine as a unit with other substances. Thus, for example, calcium bicarbonate (one of the compounds used in some antacid pills) is written $Ca(HCO_3)_2$. This shows that the part of the substance inside the brackets reacts as a unit and the "2" outside the brackets shows the presence of two such units.

Some substances attract water molecules to themselves. To show this a dot is used. Thus the blue form of copper sulphate is written $CuSO_4.5H_2O$. In this case five molecules of water attract to one of copper sulphate. When you see the dot, you know that this water can be driven off by heating; it is part of the crystal structure.

In a reaction substances change by rearranging the combinations of atoms. The way they change is shown by using the chemical symbols, placing those that will react (the starting materials, or reactants) on the left and the products of the reaction on the right. Between the two, chemists use an arrow to show which way the reaction is occurring.

It is possible to describe a reaction in words. This gives word equations, which are given throughout this book. However, it is easier to understand what is happening by using an equation containing symbols. These are also given in many places. They are not given when the equations are very complex.

In any equation both sides balance; that is, there must be an equal number of like atoms on both sides of the arrow. When you try to write down reactions, you, too, must balance your equation; you cannot have a few atoms left over at the end!

The symbols in brackets are abbreviations for the physical state of each substance taking part, so that (*s*) is used for solid, (*l*) for liquid, (*g*) for gas and (*aq*) for an aqueous solution, that is, a solution of a substance dissolved in water.

Atoms and ions
Each sphere represents a particle of an element. A particle can be an atom or an ion. Each atom or ion is associated with other atoms or ions through bonds – forces of attraction. The size of the particles and the nature of the bonds can be extremely important in determining the nature of the reaction or the properties of the compound.

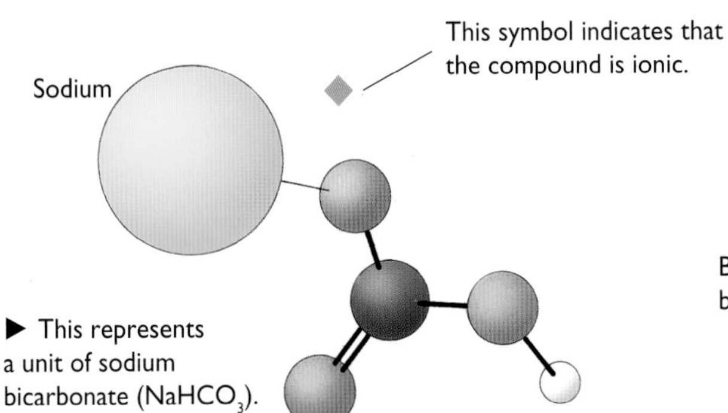

▶ This represents a unit of sodium bicarbonate ($NaHCO_3$).

The term "unit" is sometimes used to simplify the representation of a combination of ions.

Chemical symbols, equations and diagrams
The arrangement of any molecule or compound can be shown in one of the two ways shown below, depending on which gives the clearer picture. The left-hand diagram is called a ball-and-stick diagram because it uses rods and spheres to show the structure of the material. This example shows water, H_2O. There are two hydrogen atoms and one oxygen atom.

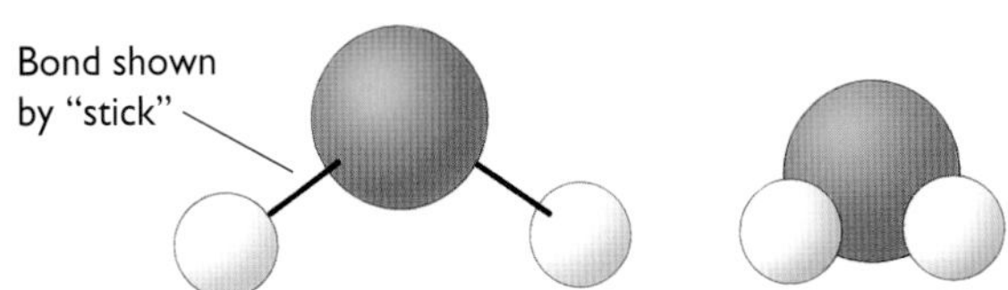

Colours too
The colours of each of the particles help differentiate the elements involved. The diagram can then be matched to the written and symbolic equation given with the diagram. In the case above, oxygen is red and hydrogen is grey.

Glossary of technical terms

absorb: to soak up a substance. Compare to adsorb.

acetone: a petroleum-based solvent.

acid: compounds containing hydrogen which can attack and dissolve many substances. Acids are described as weak or strong, dilute or concentrated, mineral or organic.

acidity: a general term for the strength of an acid in a solution.

acid rain: rain that is contaminated by acid gases such as sulphur dioxide and nitrogen oxides released by pollution.

adsorb/adsorption: to "collect" gas molecules or other particles on to the *surface* of a substance. They are not chemically combined and can be removed. (The process is called "adsorption".) Compare to absorb.

alchemy: the traditional "art" of working with chemicals that prevailed through the Middle Ages. One of the main challenges of alchemy was to make gold from lead. Alchemy faded away as scientific chemistry was developed in the 17th century.

alkali: a base in solution.

alkaline: the opposite of acidic. Alkalis are bases that dissolve, and alkaline materials are called basic materials. Solutions of alkalis have a pH greater than 7.0 because they contain relatively few hydrogen ions.

alloy: a mixture of a metal and various other elements.

alpha particle: a stable combination of two protons and two neutrons, which is ejected from the nucleus of a radioactive atom as it decays. An alpha particle is also the nucleus of the atom of helium. If it captures two electrons it can become a neutral helium atom.

amalgam: a liquid alloy of mercury with another metal.

amino acid: amino acids are organic compounds that are the building blocks for the proteins in the body.

amorphous: a solid in which the atoms are not arranged regularly (i.e. "glassy"). Compare with crystalline.

amphoteric: a metal that will react with both acids and alkalis.

anhydrous: a substance from which water has been removed by heating. Many hydrated salts are crystalline. When they are heated and the water is driven off, the material changes to an anhydrous powder.

anion: a negatively charged atom or group of atoms.

anode: the negative terminal of a battery or the positive electrode of an electrolysis cell.

anodising: a process that uses the effect of electrolysis to make a surface corrosion-resistant.

antacid: a common name for any compound that reacts with stomach acid to neutralise it.

antioxidant: a substance that prevents oxidation of some other substance.

aqueous: a solid dissolved in water. Usually used as "aqueous solution".

atom: the smallest particle of an element.

atomic number: the number of electrons or the number of protons in an atom.

atomised: broken up into a very fine mist. The term is used in connection with sprays and engine fuel systems.

aurora: the "northern lights" and "southern lights" that show as coloured bands of light in the night sky at high latitudes. They are associated with the way cosmic rays interact with oxygen and nitrogen in the air.

basalt: an igneous rock with a low proportion of silica (usually below 55%). It has microscopically small crystals.

base: a compound that may be soapy to the touch and that can react with an acid in water to form a salt and water.

battery: a series of electrochemical cells.

bauxite: an ore of aluminium, of which about half is aluminium oxide.

becquerel: a unit of radiation equal to one nuclear disintegration per second.

beta particle: a form of radiation in which electrons are emitted from an atom as the nucleus breaks down.

bleach: a substance that removes stains from materials either by oxidising or reducing the staining compound.

boiling point: the temperature at which a liquid boils, changing from a liquid to a gas.

bond: chemical bonding is either a transfer or sharing of electrons by two or more atoms. There are a number of types of chemical bond, some very strong (such as covalent bonds), others weak (such as hydrogen bonds). Chemical bonds form because the linked molecule is more stable than the unlinked atoms from which it formed. For example, the hydrogen molecule (H_2) is more stable than single atoms of hydrogen, which is why hydrogen gas is always found as molecules of two hydrogen atoms.

brass: a metal alloy principally of copper and zinc.

brazing: a form of soldering, in which brass is used as the joining metal.

brine: a solution of salt (sodium chloride) in water.

bronze: an alloy principally of copper and tin.

buffer: a chemistry term meaning a mixture of substances in solution that resists a change in the acidity or alkalinity of the solution.

capillary action: the tendency of a liquid to be sucked into small spaces, such as between objects and through narrow-pore tubes. The force to do this comes from surface tension.

catalyst: a substance that speeds up a chemical reaction but itself remains unaltered at the end of the reaction.

cathode: the positive terminal of a battery or the negative electrode of an electrolysis cell.

cathodic protection: the technique of making the object that is to be protected from corrosion into the cathode of a cell. For example, a material, such as steel, is protected by coupling it with a more reactive metal, such as magnesium. Steel forms the cathode and magnesium the anode. Zinc protects steel in the same way.

cation: a positively charged atom or group of atoms.

caustic: a substance that can cause burns if it touches the skin.

cell: a vessel containing two electrodes and an electrolyte that can act as an electrical conductor.

ceramic: a material based on clay minerals, which has been heated so that it has chemically hardened.

chalk: a pure form of calcium carbonate made of the crushed bodies of microscopic sea creatures, such as plankton and algae.

change of state: a change between one of the three states of matter, solid, liquid and gas.

chlorination: adding chlorine to a substance.

cladding: a surface sheet of material designed to protect other materials from corrosion.

clay: a microscopically small plate-like mineral that makes up the bulk of many soils. It has a sticky feel when wet.

combustion: the special case of oxidisation of a substance where a considerable amount of heat and usually light are given out. Combustion is often referred to as "burning".

compound: a chemical consisting of two or more elements chemically bonded together. Calcium atoms can combine with carbon atoms and oxygen atoms to make calcium carbonate, a compound of all three atoms.

condensation nuclei: microscopic particles of dust, salt and other materials suspended in the air, which attract water molecules.

conduction: (i) the exchange of heat (heat conduction) by contact with another object or (ii) allowing the flow of electrons (electrical conduction).

convection: the exchange of heat energy with the surroundings produced by the flow of a fluid due to being heated or cooled.

corrosion: the *slow* decay of a substance resulting from contact with gases and liquids in the environment. The term is often applied to metals. Rust is the corrosion of iron.

corrosive: a substance, either an acid or an alkali, that *rapidly* attacks a wide range of other substances.

cosmic rays: particles that fly through space and bombard all atoms on the Earth's surface. When they interact with the atmosphere they produce showers of secondary particles.

covalent bond: the most common form of strong chemical bonding, which occurs when two atoms *share* electrons.

cracking: breaking down complex molecules into simpler components. It is a term particularly used in oil refining.

crude oil: a chemical mixture of petroleum liquids. Crude oil forms the raw material for an oil refinery.

crystal: a substance that has grown freely so that it can develop external faces. Compare with crystalline, where the atoms are not free to form individual crystals and amorphous where the atoms are arranged irregularly.

crystalline: the organisation of atoms into a rigid "honeycomb-like" pattern without distinct crystal faces.

crystal systems: seven patterns or systems into which all of the world's crystals can be grouped. They are: cubic, hexagonal, rhombohedral, tetragonal, orthorhombic, monoclinic and triclinic.

cubic crystal system: groupings of crystals that look like cubes.

curie: a unit of radiation. The amount of radiation emitted by 1 g of radium each second. (The curie is equal to 37 billion becquerels.)

current: an electric current is produced by a flow of electrons through a conducting solid or ions through a conducting liquid.

decay (radioactive decay): the way that a radioactive element changes into another element decause of loss of mass through radiation. For example uranium decays (changes) to lead.

decompose: to break down a substance (for example by heat or with the aid of a catalyst) into simpler components. In such a chemical reaction only one substance is involved.

dehydration: the removal of water from a substance by heating it, placing it in a dry atmosphere, or through the action of a drying agent.

density: the mass per unit volume (e.g. g/cc).

desertification: a process whereby a soil is allowed to become degraded to a state in which crops can no longer grow, i.e. desert-like. Chemical desertification is usually the result of contamination with halides because of poor irrigation practices.

detergent: a petroleum-based chemical that removes dirt.

diaphragm: a semipermeable membrane – a kind of ultra-fine mesh filter – that will allow only small ions to pass through. It is used in the electrolysis of brine.

diffusion: the slow mixing of one substance with another until the two substances are evenly mixed.

digestive tract: the system of the body that forms the pathway for food and its waste products. It begins at the mouth and includes the stomach and the intestines.

dilute acid: an acid whose concentration has been reduced by a large proportion of water.

diode: a semiconducting device that allows an electric current to flow in only one direction.

disinfectant: a chemical that kills bacteria and other microorganisms.

dissociate: to break apart. In the case of acids it means to break up forming hydrogen ions. This is an example of ionisation. Strong acids dissociate completely. Weak acids are not completely ionised and a solution of a weak acid has a relatively low concentration of hydrogen ions.

dissolve: to break down a substance in a solution without a resultant reaction.

distillation: the process of separating mixtures by condensing the vapours through cooling.

doping: adding metal atoms to a region of silicon to make it semiconducting.

dye: a coloured substance that will stick to another substance, so that both appear coloured.

electrode: a conductor that forms one terminal of a cell.

electrolysis: an electrical–chemical process that uses an electric current to cause the break up of a compound and the movement of metal ions in a solution. The process happens in many natural situations (as for example in rusting) and is also commonly used in industry for purifying (refining) metals or for plating metal objects with a fine, even metal coating.

electrolyte: a solution that conducts electricity.

electron: a tiny, negatively charged particle that is part of an atom. The flow of electrons through a solid material such as a wire produces an electric current.

electroplating: depositing a thin layer of a metal onto the surface of another substance using electrolysis.

element: a substance that cannot be decomposed into simpler substances by chemical means

emulsion: tiny droplets of one substance dispersed in another. A common oil in water emulsion is milk. The tiny droplets in an emulsion tend to come together, so another stabilising substance is often needed to wrap the particles of grease and oil in a stable coat. Soaps and detergents are such agents. Photographic film is an example of a solid emulsion.

endothermic reaction: a reaction that takes heat from the surroundings. The reaction of carbon monoxide with a metal oxide is an example.

enzyme: organic catalysts in the form of proteins in the body that speed up chemical reactions. Every living cell contains hundreds of enzymes, which ensure that the processes of life continue. Should enzymes be made inoperative, such as through mercury poisoning, then death follows.

ester: organic compounds, formed by the reaction of an alcohol with an acid, which often have a fruity taste.

evaporation: the change of state of a liquid to a gas. Evaporation happens below the boiling point and is used as a method of separating out the materials in a solution.

exothermic reaction: a reaction that gives heat to the surroundings. Many oxidation reactions, for example, give out heat.

explosive: a substance which, when a shock is applied to it, decomposes very rapidly, releasing a very large amount of heat and creating a large volume of gases as a shock wave.

extrusion: forming a shape by pushing it through a die. For example, toothpaste is extruded through the cap (die) of the toothpaste tube.

fallout: radioactive particles that reach the ground from radioactive materials in the atmosphere.

fat: semi-solid energy-rich compounds derived from plants or animals and which are made of carbon, hydrogen and oxygen. Scientists call these esters.

feldspar: a mineral consisting of sheets of aluminium silicate. This is the mineral from which the clay in soils is made.

fertile: able to provide the nutrients needed for unrestricted plant growth.

filtration: the separation of a liquid from a solid using a membrane with small holes.

fission: the breakdown of the structure of an atom, popularly called "splitting the atom" because the atom is split into approximately two other nuclei. This is different from, for example, the small change that happens when radioactivity is emitted.

fixation of nitrogen: the processes that natural organisms, such as bacteria, use to turn the nitrogen of the air into ammonium compounds.

fixing: making solid and liquid nitrogen-containing compounds from nitrogen gas. The compounds that are formed can be used as fertilisers.

fluid: able to flow; either a liquid or a gas.

fluorescent: a substance that gives out visible light when struck by invisible waves such as ultraviolet rays.

flux: a material used to make it easier for a liquid to flow. A flux dissolves metal oxides and so prevents a metal from oxidising while being heated.

foam: a substance that is sufficiently gelatinous to be able to contain bubbles of gas. The gas bulks up the substance, making it behave as though it were semi-rigid.

fossil fuels: hydrocarbon compounds that have been formed from buried plant and animal remains. High pressures and temperatures lasting over millions of years are required. The fossil fuels are coal, oil and natural gas.

fraction: a group of similar components of a mixture. In the petroleum industry the light fractions of crude oil are those with the smallest molecules, while the medium and heavy fractions have larger molecules.

free radical: a very reactive atom or group with a "spare" electron.

freezing point: the temperature at which a substance changes from a liquid to a solid. It is the same temperature as the melting point.

fuel: a concentrated form of chemical energy. The main sources of fuels (called fossil fuels because they were formed by geological processes) are coal, crude oil and natural gas. Products include methane, propane and gasoline. The fuel for stars and space vehicles is hydrogen.

fuel rods: rods of uranium or other radioactive material used as a fuel in nuclear power stations.

fuming: an unstable liquid that gives off a gas. Very concentrated acid solutions are often fuming solutions.

fungicide: any chemical that is designed to kill fungi and control the spread of fungal spores.

fusion: combining atoms to form a heavier atom.

galvanising: applying a thin zinc coating to protect another metal.

gamma rays: waves of radiation produced as the nucleus of a radioactive element rearranges itself into a tighter cluster of protons and neutrons. Gamma rays carry enough energy to damage living cells.

gangue: the unwanted material in an ore.

gas: a form of matter in which the molecules form no definite shape and are free to move about to fill any vessel they are put in.

gelatinous: a term meaning made with water. Because a gelatinous precipitate is mostly water, it is of a similar density to water and will float or lie suspended in the liquid.

gelling agent: a semi-solid jelly-like substance.

gemstone: a wide range of minerals valued by people, both as crystals (such as emerald) and as decorative stones (such as agate). There is no single chemical formula for a gemstone.

glass: a transparent silicate without any crystal growth. It has a glassy lustre and breaks with a curved fracture. Note that some minerals have all these features and are therefore natural glasses. Household glass is a synthetic silicate.

glucose: the most common of the natural sugars. It occurs as the polymer known as cellulose, the fibre in plants. Starch is also a form of glucose. The breakdown of glucose provides the energy that animals need for life.

granite: an igneous rock with a high proportion of silica (usually over 65%). It has well-developed large crystals. The largest pink, grey or white crystals are feldspar.

Greenhouse Effect: an increase of the global air temperature as a result of heat released from burning fossil fuels being absorbed by carbon dioxide in the atmosphere.

gypsum: the name for calcium sulphate. It is commonly found as Plaster of Paris and wallboards.

half-life: the time it takes for the radiation coming from a sample of a radioactive element to decrease by half.

halide: a salt of one of the halogens (fluorine, chlorine, bromine and iodine).

halite: the mineral made of sodium chloride.

halogen: one of a group of elements including chlorine, bromine, iodine and fluorine.

heat-producing: see exothermic reaction.

high explosive: a form of explosive that will only work when it receives a shock from another explosive. High explosives are much more powerful than ordinary explosives. Gunpowder is not a high explosive.

hydrate: a solid compound in crystalline form that contains molecular water. Hydrates commonly form when a solution of a soluble salt is evaporated. The water that forms part of a hydrate crystal is known as the "water of crystallization". It can usually be removed by heating, leaving an anhydrous salt.

hydration: the absorption of water by a substance. Hydrated materials are not "wet" but remain firm, apparently dry, solids. In some cases, hydration makes the substance change colour, in many other cases there is no colour change, simply a change in volume.

hydrocarbon: a compound in which only hydrogen and carbon atoms are present. Most fuels are hydrocarbons, as is the simple plastic polyethene (known as polythene).

hydrogen bond: a type of attractive force that holds one molecule to another. It is one of the weaker forms of intermolecular attractive force.

hydrothermal: a process in which hot water is involved. It is usually used in the context of rock formation because hot water and other fluids sent outwards from liquid magmas are important carriers of metals and the minerals that form gemstones.

igneous rock: a rock that has solidified from molten rock, either volcanic lava on the Earth's surface or magma deep underground. In either case the rock develops a network of interlocking crystals.

incendiary: a substance designed to cause burning.

indicator: a substance or mixture of substances that change colour with acidity or alkalinity.

inert: nonreactive.

infra-red radiation: a form of light radiation where the wavelength of the waves is slightly longer than visible light. Most heat radiation is in the infra-red band.

insoluble: a substance that will not dissolve.

ion: an atom, or group of atoms, that has gained or lost one or more electrons and so developed an electrical charge. Ions behave differently from electrically neutral atoms and molecules. They can move in an electric field, and they can also bind strongly to solvent molecules such as water. Positively charged ions are called cations; negatively charged ions are called anions. Ions carry electrical current through solutions.

ionic bond: the form of bonding that occurs between two ions when the ions have opposite charges. Sodium cations bond with chloride anions to form common salt (NaCl) when a salty solution is evaporated. Ionic bonds are strong bonds except in the presence of a solvent.

ionise: to break up neutral molecules into oppositely charged ions or to convert atoms into ions by the loss of electrons.

ionisation: a process that creates ions.

irrigation: the application of water to fields to help plants grow during times when natural rainfall is sparse.

isotope: atoms that have the same number of protons in their nucleus, but which have different masses; for example, carbon-12 and carbon-14.

latent heat: the amount of heat that is absorbed or released during the process of changing state between gas, liquid or solid. For example, heat is absorbed when a substance melts and it is released again when the substance solidifies.

latex: (the Latin word for "liquid") a suspension of small polymer particles in water. The rubber that flows from a rubber tree is a natural latex. Some synthetic polymers are made as latexes, allowing polymerisation to take place in water.

lava: the material that flows from a volcano.

limestone: a form of calcium carbonate rock that is often formed of lime mud. Most limestones are light grey and have abundant fossils.

liquid: a form of matter that has a fixed volume but no fixed shape.

lode: a deposit in which a number of veins of a metal found close together.

lustre: the shininess of a substance.

magma: the molten rock that forms a balloon-shaped chamber in the rock below a volcano. It is fed by rock moving upwards from below the crust.

marble: a form of limestone that has been "baked" while deep inside mountains. This has caused the limestone to melt and reform into small interlocking crystals, making marble harder than limestone.

mass: the amount of matter in an object. In everyday use, the word weight is often used to mean mass.

melting point: the temperature at which a substance changes state from a solid to a liquid. It is the same as freezing point.

membrane: a thin flexible sheet. A semipermeable membrane has microscopic holes of a size that will selectively allow some ions and molecules to pass through but hold others back. It thus acts as a kind of sieve.

meniscus: the curved surface of a liquid that forms when it rises in a small bore, or capillary tube. The meniscus is convex (bulges upwards) for mercury and is concave (sags downwards) for water.

metal: a substance with a lustre, the ability to conduct heat and electricity and which is not brittle.

metallic bonding: a kind of bonding in which atoms reside in a "sea" of mobile electrons. This type of bonding allows metals to be good conductors and means that they are not brittle

metamorphic rock: formed either from igneous or sedimentary rocks, by heat and or pressure. Metamorphic rocks form deep inside mountains during periods of mountain building. They result from the remelting of rocks during which process crystals are able to grow. Metamorphic rocks often show signs of banding and partial melting.

micronutrient: an element that the body requires in small amounts. Another term is trace element.

mineral: a solid substance made of just one element or chemical compound. Calcite is a mineral because it consists only of calcium carbonate, halite is a mineral because it contains only sodium chloride, quartz is a mineral because it consists of only silicon dioxide.

mineral acid: an acid that does not contain carbon and that attacks minerals. Hydrochloric, sulphuric and nitric acids are the main mineral acids.

mineral-laden: a solution close to saturation.

mixture: a material that can be separated out into two or more substances using physical means.

molecule: a group of two or more atoms held together by chemical bonds.

monoclinic system: a grouping of crystals that look like double-ended chisel blades.

monomer: a building block of a larger chain molecule ("mono" means one, "mer" means part).

mordant: any chemical that allows dyes to stick to other substances.

native metal: a pure form of a metal, not combined as a compound. Native metal is more common in poorly reactive elements than in those that are very reactive.

neutralisation: the reaction of acids and bases to produce a salt and water. The reaction causes hydrogen from the acid and hydroxide from the base to be changed to water. For example, hydrochloric acid reacts with sodium hydroxide to form common salt and water. The term is more generally used for any reaction where the pH changes towards 7.0, which is the pH of a neutral solution.

neutron: a particle inside the nucleus of an atom that is neutral and has no charge.

noncombustible: a substance that will not burn.

noble metal: silver, gold, platinum, and mercury. These are the least reactive metals.

nuclear energy: the heat energy produced as part of the changes that take place in the core, or nucleus, of an element's atoms.

nuclear reactions: reactions that occur in the core, or nucleus of an atom.

nutrients: soluble ions that are essential to life.

octane: one of the substances contained in fuel.

ore: a rock containing enough of a useful substance to make mining it worthwhile.

organic acid: an acid containing carbon and hydrogen.

organic substance: a substance that contains carbon.

osmosis: a process where molecules of a liquid solvent move through a membrane (filter) from a region of low concentration to a region of high concentration of solute.

oxidation: a reaction in which the oxidising agent removes electrons. (Note that oxidising agents do not have to contain oxygen.)

oxide: a compound that includes oxygen and one other element.

oxidise: the process of gaining oxygen. This can be part of a controlled chemical reaction, or it can be the result of exposing a substance to the air, where oxidation (a form of corrosion) will occur slowly, perhaps over months or years.

oxidising agent: a substance that removes electrons from another substance (and therefore is itself reduced).

ozone: a form of oxygen whose molecules contain three atoms of oxygen. Ozone is regarded as a beneficial gas when high in the atmosphere because it blocks ultraviolet rays. It is a harmful gas when breathed in, so low level ozone, which is produced as part of city smog, is regarded as a form of pollution. The ozone layer is the uppermost part of the stratosphere.

pan: the name given to a shallow pond of liquid. Pans are mainly used for separating solutions by evaporation.

patina: a surface coating that develops on metals and protects them from further corrosion.

percolate: to move slowly through the pores of a rock.

period: a row in the Periodic Table.

Periodic Table: a chart organising elements by atomic number and chemical properties into groups and periods.

pesticide: any chemical that is designed to control pests (unwanted organisms) that are harmful to plants or animals.

petroleum: a natural mixture of a range of gases, liquids and solids derived from the decomposed remains of plants and animals.

pH: a measure of the hydrogen ion concentration in a liquid. Neutral is pH 7.0; numbers greater than this are alkaline, smaller numbers are acidic.

phosphor: any material that glows when energized by ultraviolet or electron beams such as in fluorescent tubes and cathode ray tubes. Phosphors, such as phosphorus, emit light after the source of excitation is cut off. This is why they glow in the dark. By contrast, fluorescors, such as fluorite, emit light only while they are being excited by ultraviolet light or an electron beam.

photon: a parcel of light energy.

photosynthesis: the process by which plants use the energy of the Sun to make the compounds they need for life. In photosynthesis, six molecules of carbon dioxide from the air combine with six molecules of water, forming one molecule of glucose (sugar) and releasing six molecules of oxygen back into the atmosphere.

pigment: any solid material used to give a liquid a colour.

placer deposit: a kind of ore body made of a sediment that contains fragments of gold ore eroded from a mother lode and transported by rivers and/or ocean currents.

plastic (material): a carbon-based material consisting of long chains (polymers) of simple molecules. The word plastic is commonly restricted to synthetic polymers.

plastic (property): a material is plastic if it can be made to change shape easily. Plastic materials will remain in the new shape. (Compare with elastic, a property where a material goes back to its original shape.)

plating: adding a thin coat of one material to another to make it resistant to corrosion.

playa: a dried-up lake bed that is covered with salt deposits. From the Spanish word for beach.

poison gas: a form of gas that is used intentionally to produce widespread injury and death. (Many gases are poisonous, which is why many chemical reactions are performed in laboratory fume chambers, but they are a byproduct of a reaction and not intended to cause harm.)

polymer: a compound that is made of long chains by combining molecules (called monomers) as repeating units. ("Poly" means many, "mer" means part).

polymerisation: a chemical reaction in which large numbers of similar molecules arrange themselves into large molecules, usually long chains. This process usually happens when there is a suitable catalyst present. For example, ethene reacts to form polythene in the presence of certain catalysts.

porous: a material containing many small holes or cracks. Quite often the pores are connected, and liquids, such as water or oil, can move through them.

precious metal: silver, gold, platinum, iridium, and palladium. Each is prized for its rarity. This category is the equivalent of precious stones, or gemstones, for minerals.

precipitate: tiny solid particles formed as a result of a chemical reaction between two liquids or gases.

preservative: a substance that prevents the natural organic decay processes from occurring. Many substances can be used safely for this purpose, including sulphites and nitrogen gas.

product: a substance produced by a chemical reaction.

protein: molecules that help to build tissue and bone and therefore make new body cells. Proteins contain amino acids.

proton: a positively charged particle in the nucleus of an atom that balances out the charge of the surrounding electrons

pyrite: "mineral of fire". This name comes from the fact that pyrite (iron sulphide) will give off sparks if struck with a stone.

pyrometallurgy: refining a metal from its ore using heat. A blast furnace or smelter is the main equipment used.

radiation: the exchange of energy with the surroundings through the transmission of waves or particles of energy. Radiation is a form of energy transfer that can happen through space; no intervening medium is required (as would be the case for conduction and convection).

radioactive: a material that emits radiation or particles from the nucleus of its atoms.

radioactive decay: a change in a radioactive element due to loss of mass through radiation. For example uranium decays (changes) to lead.

radioisotope: a shortened version of the phrase radioactive isotope.

radiotracer: a radioactive isotope that is added to a stable, nonradioactive material in order to trace how it moves and its concentration.

reaction: the recombination of two substances using parts of each substance to produce new substances.

reactivity: the tendency of a substance to react with other substances. The term is most widely used in comparing the reactivity of metals. Metals are arranged in a reactivity series.

reagent: a starting material for a reaction.

recycling: the reuse of a material to save the time and energy required to extract new material from the Earth and to conserve non-renewable resources.

redox reaction: a reaction that involves reduction and oxidation.

reducing agent: a substance that gives electrons to another substance. Carbon monoxide is a reducing agent when passed over copper oxide, turning it to copper and producing carbon dioxide gas. Similarly, iron oxide is reduced to iron in a blast furnace. Sulphur dioxide is a reducing agent, used for bleaching bread.

reduction: the removal of oxygen from a substance. See also: oxidation.

refining: separating a mixture into the simpler substances of which it is made. In the case of a rock, it means the extraction of the metal that is mixed up in the rock. In the case of oil it means separating out the fractions of which it is made.

refractive index: the property of a transparent material that controls the angle at which total internal reflection will occur. The greater the refractive index, the more reflective the material will be.

resin: natural or synthetic polymers that can be moulded into solid objects or spun into thread.

rust: the corrosion of iron and steel.

saline: a solution in which most of the dissolved matter is sodium chloride (common salt).

salinisation: the concentration of salts, especially sodium chloride, in the upper layers of a soil due to poor methods of irrigation.

salts: compounds, often involving a metal, that are the reaction products of acids and bases. (Note "salt" is also the common word for sodium chloride, common salt or table salt.)

saponification: the term for a reaction between a fat and a base that produces a soap.

saturated: a state where a liquid can hold no more of a substance. If any more of the substance is added, it will not dissolve.

saturated solution: a solution that holds the maximum possible amount of dissolved material. The amount of material in solution varies with the temperature; cold solutions can hold less dissolved solid material than hot solutions. Gases are more soluble in cold liquids than hot liquids.

sediment: material that settles out at the bottom of a liquid when it is still.

semiconductor: a material of intermediate conductivity. Semiconductor devices often use silicon when they are made as part of diodes, transistors or integrated circuits.

semipermeable membrane: a thin (membrane) of material that acts as a fine sieve, allowing small molecules to pass, but holding large molecules back.

silicate: a compound containing silicon and oxygen (known as silica).

sintering: a process that happens at moderately high temperatures in some compounds. Grains begin to fuse together even through they do not melt. The most widespread example of sintering happens during the firing of clays to make ceramics.

slag: a mixture of substances that are waste products of a furnace. Most slags are composed mainly of silicates.

smelting: roasting a substance in order to extract the metal contained in it.

smog: a mixture of smoke and fog. The term is used to describe city fogs in which there is a large proportion of particulate matter (tiny pieces of carbon from exhausts) and also a high concentration of sulphur and nitrogen gases and probably ozone.

soldering: joining together two pieces of metal using solder, an alloy with a low melting point.

solid: a form of matter where a substance has a definite shape.

soluble: a substance that will readily dissolve in a solvent.

solute: the substance that dissolves in a solution (e.g. sodium chloride in salt water).

solution: a mixture of a liquid and at least one other substance (e.g. salt water). Mixtures can be separated out by physical means, for example by evaporation and cooling.

solvent: the main substance in a solution (e.g. water in salt water).

spontaneous combustion: the effect of a very reactive material beginning to oxidise very quickly and bursting into flame.

stable: able to exist without changing into another substance.

stratosphere: the part of the Earth's atmosphere that lies immediately above the region in which clouds form. It occurs between 12 and 50 km above the Earth's surface.

strong acid: an acid that has completely dissociated (ionised) in water. Mineral acids are strong acids.

sublimation: the change of a substance from solid to gas, or vica versa, without going through a liquid phase.

substance: a type of material, including mixtures.

sulphate: a compound that includes sulphur and oxygen, for example, calcium sulphate or gypsum.

sulphide: a sulphur compound that contains no oxygen.

sulphite: a sulphur compound that contains less oxygen than a sulphate.

surface tension: the force that operates on the surface of a liquid, which makes it act as though it were covered with an invisible elastic film.

suspension: tiny particles suspended in a liquid.

synthetic: does not occur naturally, but has to be manufactured.

tarnish: a coating that develops as a result of the reaction between a metal and substances in the air. The most common form of tarnishing is a very thin transparent oxide coating.

thermonuclear reactions: reactions that occur within atoms due to fusion, releasing an immensely concentrated amount of energy.

thermoplastic: a plastic that will soften, can repeatedly be moulded it into shape on heating and will set into the moulded shape as it cools.

thermoset: a plastic that will set into a moulded shape as it cools, but which cannot be made soft by reheating.

titration: a process of dripping one liquid into another in order to find out the amount needed to cause a neutral solution. An indicator is used to signal change.

toxic: poisonous enough to cause death.

translucent: almost transparent.

transmutation: the change of one element into another.

vapour: the gaseous form of a substance that is normally a liquid. For example, water vapour is the gaseous form of liquid water.

vein: a mineral deposit different from, and usually cutting across, the surrounding rocks. Most mineral and metal-bearing veins are deposits filling fractures. The veins were filled by hot, mineral-rich waters rising upwards from liquid volcanic magma. They are important sources of many metals, such as silver and gold, and also minerals such as gemstones. Veins are usually narrow, and were best suited to hand-mining. They are less exploited in the modern machine age.

viscous: slow moving, syrupy. A liquid that has a low viscosity is said to be mobile.

vitreous: glass-like.

volatile: readily forms a gas.

vulcanisation: forming cross-links between polymer chains to increase the strength of the whole polymer. Rubbers are vulcanised using sulphur when making tyres and other strong materials.

weak acid: an acid that has only partly dissociated (ionised) in water. Most organic acids are weak acids.

weather: a term used by Earth scientists and derived from "weathering", meaning to react with water and gases of the environment.

weathering: the slow natural processes that break down rocks and reduce them to small fragments either by mechanical or chemical means.

welding: fusing two pieces of metal together using heat.

X-rays: a form of very short wave radiation.

Index